THE CRISIS IN WESTERN SECURITY

THE CRISIS IN WESTERN SECURITY

EDITED BY LAWRENCE S. HAGEN

ST. MARTIN'S PRESS NEW YORK

 For information, write:
St. Martin's Press, Inc., 175 Fifth Avenue, New York, NY 10010
Printed in Great Britain
First published in the United States of America in 1982

Library of Congress Cataloging in Publication Data
Main entry under title:

The Crisis in Western security.

1. North Atlantic Treaty Organization – Addresses, essays, lectures. I. Hagen, Lawrence S.
UA646.3.C74 355'.031'091821 81-14340
ISBN 0-312-17397-0

Printed and bound in Great Britain

CONTENTS

Acknowledgements

Introduction: Contemporary Strategic Relations and the Dissolution of Detente *Lawrence S. Hagen* 9

1. On the Logic of Security and Arms Control in the NATO Alliance *Philip Windsor* 27

2. Arms Control: The Possibility of a Second Coming *Lawrence Freedman* 41

3. Arms Control and Western Security: A Question of Growing Irrelevance *Richard Burt* 56

4. Nuclear Arms Control and Europe: The Enduring Dilemma *Robin Ranger* 74

5. Arms Control and European Security: Some Basic Issues *Colin S. Gray* 94

6. The Future of the Strategic Balance *Desmond J. Ball* 121

7. Technology, Deterrence and the NATO Alliance *Lothar Ruehl* 144

8. NATO and Long-range Theatre Nuclear Weapons: Background and Rationale *Paul Buteux* 152

9. Who is Decoupling From Whom? *Or* This Time, the Wolf is Here *Pierre Hassner* 168

10. East-West Relations and the Politics of Security *Helmut Sonnenfeldt* 186

11. Military Power and Arms Control: Towards a Reassessment *Hugh Macdonald* 196

Notes on Contributors 242

Index 244

Contents

Acknowledgements

Introduction [illegible]

[illegible]

Notes on Contributors

Index

ACKNOWLEDGEMENTS

This volume owes its origins to a recent conference on arms control and Western security sponsored by *Millennium: Journal of International Studies* at the London School of Economics. For their financial support of that event, I would like to thank the Ford Foundation, the International Centre for Economics and Related Disciplines at the LSE, and the United States Embassy in London. The Editors of *Millennium* (Jana Bennett, Anne Rolfe, Jane Sargent and Joyce Colwill) and the Editorial Board deserve particular thanks for their organisational assistance and moral support, as do the Department of International Relations and the School itself. For his intellectual stimulation and encouragement, I would like to express my deep personal gratitude to Philip Windsor. Most of all, I would like to thank Hilary Parker, Joyce Colwill, Robert Fonow and Martin McCusker; their energy and assistance in helping to see this project through to fruition was truly second to none.

LSH
London

INTRODUCTION: CONTEMPORARY STRATEGIC RELATIONS AND THE DISSOLUTION OF DETENTE

Lawrence S. Hagen

Humpty Dumpty:	When I use a word it means just what I choose it to mean – neither more nor less.
Alice:	The question is whether you *can* make words mean so many different things.

(*Through the Looking Glass, Chapter 6*)

One of the most enduring insights to be derived from Clausewitz concerns the criterion for determining military victory. Success in war does not lie, in the first instance, with the battlefield. Rather, the purpose of intelligent battle is to engineer a context in which the opponent agrees to abide by the rules of political intercourse as defined by the victor. In the absence of such agreement, battlefield supremacy is likely to be Pyrrhic, merely defining the terms of the next conflict. In this sense, military success is consequential, not causal; the opponent *agrees* to lose. Disregard for this lesson peppers history, with the Franco-Prussian War, the First World War, and the Arab-Israeli War of 1967 – and 1973 – providing examples. In each case, the question of an acceptable international or regional order, and the proper scope for change within it, was not resolved by the process of war; indeed, that process had as its effect the creation of new conflicts which transcended the issues which formed the basis for hostility in the first place.

The period since the Second World War has proven to be no exception to this process; subsequent political conflicts between victors, and between victors and vanquished, have provided the central dynamic of strategic and political relations between the participating states of that war. Moreover, as with previous unresolved hostilities, the post-war era has been littered with attempts by the relevant actors within it to define an acceptable context of order, and to articulate and institutionalise criteria for the determination of legitimate modes and types of change – in other words, to lend predictability to the international system in the absence of a clear consensus as to the nature of security.

Initially, this process was informed by a sort of strategic despair. The collapse of Western, but primarily American, assumptions as to the preferred structure for post-war international order – based on US-

Soviet co-operation – gave way to a perceived vacuum which, when compounded with the disillusionment and suspicion which this 'betrayal' generated, fed into the subjectification – and subsequent objectification of what we now call the Cold War. A stillborn mechanism of order was replaced by a surrogate system predicated upon a symbiosis: on the one hand, there existed the hostility between the Soviet Union and the United States, and, on the other, the co-operative framework of Western states in the form of an informal – and subsequently formal – alliance of interests based on consensual values. A 'tight bipolarity' would provide a negative order by creating and maintaining geographical and behavioural boundaries which defined the scope of legitimate Superpower activity; the Western Alliance would, in a more positive sense, both effect that bipolarity through the creation of a strong system of mutual prosperity and defence, and provide for a mechanism for the exercise of control in the international system as a whole. Acting as a meta-concept for providing the criteria of change and constancy was the policy of containment.

But this order, characterised by the transcendence of political and economic conflict in a context of generalised unpredictability, and the emergence of a system of security defined largely in military terms, had sown within it the seeds of its own transformation. For to the extent that the military confrontation reached a set of stable configurations, and a clear understanding of the *purposes* served by military force was created through challenge and response, threat and reassurance, one could begin to talk not simply about *order,* but about legitimate *change* within the stable context created by the working through of the security relationship. As one observer notes, 'a Cold War can be avoided when the real areas of contention are clear, and the consequences of change predictable – where it is clear what can be the subject of bargaining and what cannot be'[1]: an order is posited whereby that order is challenged.

The period of post-war international relations which we call detente has been the direct consequence of this dialectical process of the creation of order and subsequent questioning of that order. But this attempt at the creation of a new order has failed; detente has created a *disorder* according to which the old order has been re-established. It is to this process of dissolution which this study is directed.

This assertion immediately raises problems of definition. By what criteria does one judge 'failure', and by what characteristics does one define 'the Cold War' and its progeny, 'detente'? To a large extent, such questions are sterile, to be answered by the process of enquiry rather than through a sort of definitional deductivism. Nevertheless, two observations bear delineation at this point. First, according to the meaning given to it by its participants – the only legitimate criterion for success – detente has manifestly failed either to provide for a significant and lasting reduction in Cold War tensions, or to erect a new mechanism for East-West order. On this point there can be little argument, only debate as to the degree or permanence of failure. Secondly, it must be noted that those who assert, more through hope than analysis, the continued operation of a meaningful detente simply because we have yet to return to the hysteria and insecurity of the worst years of the Cold War, must necessarily embrace a rigidly circular philosophy of history. To claim that the present is not like the past is to do nothing more than acknowledge the forward movement of history; it does little to describe or explain its machinations or direction.

A second proposition, one which permeates this volume, is less one of history than of historical interpretation: that the processes which have fed into the dissolution of detente have been *intrinsic* to the phenomenon itself, and not the result of contingent tactical error or faulty judgement, although there has been enough of this. A sort of strategic entropy born of the interconnections between the intellectual premisses of those who have participated in detente, the resulting policies which were designed to serve as the engines of detente, and the structural context within which these policies have operated, has acted to dissolve detente from within. This is not, of course, to argue in favour of an operant determinism. The connections assumed, and drawn, between various processes of East-West relations have been the result of particular sets of human decisions, and therefore expressions of choice; they are the product of the strategic 'mind' brought to bear on strategic 'matter', and as such enjoy a relative autonomy. What *is* asserted, however, is that given the context of detente, the particular choices made, and the results of these choices in their manifestations in policy, detente has been *bestowed* with an incipient auto-destabilisation from which it has not been able to escape.

Detente, then, is to be seen not simply as a failure of practice, but as a failure of conception. The Republic of Strategists, whose distilled wisdom came to inform American policy, has disappeared with the policy to which it gave its *imprimatur*; the contours of strategic discourse, from the basic assumptions of deterrence through to institutional choreographies of operationalisation, have been thrown into disarray. From an initial attempt to determine the *meaning* of nuclear weapons for the conduct of international politics, we no longer know what the criteria of meaning are. In this sense, the detente process evinces the dilemmas which have exercised strategic thinkers and actors since the advent of the bipolar nuclear age: how to arbitrate between the negative demands of deterrence and those of the management of positive change; how to reconcile the nuclear 'fact' with the independent world of 'values'; how to relate the manifest irrationality of nuclear deterrence with the requirements of rationality at other levels, and in other contexts; and how to harness the subjective implications of deterrence in a manner which allows its perpetuation through institutionalisation, and perhaps even permits of a useful extension of its stabilising ambit. To set the context for the rest of this volume, it is, perhaps, useful to examine briefly how detente has tried to answer these questions.

The Logic of Detente

Detente has its origins in an evolved geopolitical certainty in the context of an incipient strategic *un*certainty; neither in and of itself provided sufficient grounds for the transformation of the Cold War, but the synergism which resulted from the coexistence of these two trends yielded a fundamental reordering of the assumptions of the Superpowers as to the requirements of stability, and the possibilities for change in the East-West system.

As the system of security defined in military terms by the Superpowers and their respective alliances approached the limits of its logic during the 1950s and early 1960s, a certainty founded on the nature of order, and the proper scope for change, began to emerge. The succession of crises in European and Superpower spheres – Czechoslovakia, Yugoslavia, Korea and the various Berlin episodes – and the accompanying accretion and cohesion in military capability within the West and the East, engendered a confidence that short-term fears had been reduced, and that new questions about the future (not simply repetitions of questions asked in the past) could now be raised. For if 'life' had been secured, how was 'the pursuit of happiness' to be conducted? Was it now possible to alter the basis of power (in the form of military confrontation) through accepting and legitimising the past?

Yet these questions could not be approached in the absence of a revolution in *strategic* relations. For developments at this level were seen, in the context of prevailing doctrine, as extremely worrisome. The Soviet Union was challenging the American capability to engage in strategic nuclear war as a rational counter to Soviet preponderance in Europe through the development of quantities and qualities of nuclear power at the intercontinental level. The United States, for its part, appeared more willing than ever to maintain its superiority in a manner which implied strategic defeat for the Soviet Union. The USSR detected a long-term threat to its security in the policies of the Kennedy Administration while the United States, in the light of its own interests, saw a short-term threat to its own strategic policy. Precisely *because* the technological balance was being transformed, doctrines of war-fighting at different levels were simultaneously challenged and then re-asserted with enhanced vigour. Subjective change was not reflective of objective change.

Illumination was, however, at hand. For through the very attempts of both Superpowers to preserve the anachronistic logic of their respective strategies, a crisis was created which gave life to a psychological reappraisal of the basis of their strategic intercourse. The Cuban Missile Crisis provided the mechanism by which this new choice was made. By demonstrating that the Soviet capacity for interference in the American promise to go rationally to war was extant or imminent, and that there existed the potential for profound instability in such a situation, the United States came tacitly to accept what had exercised some European strategists since the late 1950s – that strategic nuclear war was no longer a rational option. The most significant revolution in American strategic thought began. Credibility was to be based not on the instrumental nature of conflict – for this was no longer credible – but upon the declaration that despite this (rational) incredibility, one was still prepared to go to war because one would become irrational in a situation of threat to central values. What this in turn seemed to imply was that both sides would respect – in a rational manner – the irrationality of the other's threat, and so long as this mutual respect survived, stability would ensue.[2]

In Clausewitzian terms, both sides had agreed to lose, and in so doing to abide by the political relations and commitments to interests which each side had come to recognise through the process of the Cold War. The consequences of this subjective reformation were profound. In a period when the United States retained a formidable war-winning capability, technical reality was ignored; MAD was embraced, and parity was *chosen* by America as the correct relationship for two powers who now recognised that each had vital interests which both could threaten, and which must therefore be respected. Geopolitical sources of conflict were subsequently removed (Berlin, for example), and in those peripheral areas where commitments and interests were unclear – as in the Middle East or Vietnam – the dangers of brinkmanship were replaced by relative acquiescence, as they were in areas of obvious connection to central values, such as Czechoslovakia.

But this psychological reaction to the Cuban Missile Crisis was soon superceded by another interpretation of the implications of an irrational deterrence system. For it was logically clear that the shift from rational defence to irrational commitment simultaneously contracted the purview of nuclear deterrence; in emphasising the necessary condition of a situation where central values were threatened, less serious challenges were excluded. An unavoidable by-product of the shift in nuclear doctrine was thus the loosening of the bipolar system; the frequent injunctions of André Beaufre to the effect that if nuclear weapons were to have an ultimately stabilising effect on international conflict, it would be necessary to preserve some salutory *in*stability at the strategic nuclear level, seem to have been vindicated.

But there was also a complementary implication. If assured destruction was to be the operating relationship at the nuclear level, then war – and the danger of war through crisis – was to be avoided. Other processes and contexts of international politics had to be segregated from the central task of preserving the structure represented by MAD; the nuclear relationship was to become the meta-system according to which all questions of value were to be approached. But this in turn implied the subordination of values to order. There is nothing *a priori* invalid or contradictory about this; historical analogies such as the Westphalian and the post-Napoleonic state systems sought – and succeeded for a while – to preserve an order which precluded the pursuit of values against the survival of the system itself. Problems arise, however, when the bilateral nuclear system is examined in the light of the nature of the deterrent relationship on which it is based. For if the commitment to the defence of values is replaced by a commitment to order which seeks to engender strategic reassurance through various procedural replacements for traditional political antagonisms, this implies a weakening of the threat to go to war in the defence of those values. In other words, it weakens the very basis of the stable bipolar relationship which formed the necessary condition for detente in the first place. This would present few problems if a system of reassurance was indeed brought into play. But if the minimal structural consensus on which stability is based cannot be extended into the wider processes of Superpower and global politics, self-encapsulation is likely to be the result; the reality of detente becomes *immobilisme*.

Evidence for this proposition may be located at several levels, but perhaps the most striking has been that of peripheral conflict. In the Cold War, participation – direct or indirect – by either Superpower in regional conflict or crises was either deterred, through the threat of rational retaliation, or else an expected mode of behaviour due to the hostility explicit within a tight bipolar system. Regional hostility, then, was an indication of larger relationships, and constituted a productive signalling device for the maintenance of that relationship, and its working through to a situation of greater predictability. However, with detente, peripheral conflict has become a *determinant* of those relationships. It is here where the *logical* implications of MAD come into play, for detente makes it easier for either Superpower, given political will, to act in lower level conflicts as a result of the contraction of the threat. Moreover, if a structure of reassurance is appended to this situation of MAD (itself a highly problematic connection), interventions look much worse, since they appear to be contrary to (one state's) interpretation of detente.

If structures of geopolitical containment remain in the presence of a structure of reassurance, there then emerges a desire to reinvoke the very threat of war which detente was designed to supplant. The dangers in this process result from the prospect that the Superpower against whom this action is directed will *not* respond (as in Angola, or the Horn of Africa) or else will ultimately reinvoke the threat at a much more dangerous level (as with the 1973 US World-Wide Military Alert during the Yom Kippur War). Moreover, if the interpretation of the nature and implications of detente is asymmetrical *vis-à-vis* the need to avoid conflicts over peripheral values, inaction by the Superpower who saw the nuclear level as *determining* rather than *detached* will produce a situation of short-term inaction, with long-term implications of destabilisation, or a short-term heightening of tensions. Thus, not only does peripheral conflict come to act against the global order, it comes also to act against the initial premiss of war-avoidance.

In addition to the most recent catalogue of peripheral 'conflicts-as-criteria' (Afghanistan, 'the Soviet Brigade' in Cuba, etc.) one may, in fact be seeing a new phenomenon, with most serious implications. For what the situation in Poland, and the threatened Western response in the case of invasion, may imply is that detente has, perhaps, even begun to erode the consensus of legitimate interests which the Cold War at its latter stages, and the early years of detente, was seen to imply. The comparison between actual Western response to the invasion of Czechoslovakia and the likely Western response to a similar action in Poland is instructive. At the very least, it indicates that the corrosion of detente as a consequence of peripheral conflict and SALT-related increases in hostility may have resulted in a situation of mutual suspicion whereby what is still seen as a 'legitimate' Soviet interest is *also* seen as a factor implying – or demanding – a suspension of the detente process. This, if nothing else, is an indicator of the parlous state of detente.

But the question of peripheral conflict raises even more fundamental questions. There is in detente something of an attempt to take refuge in the familiar contours of East-West conflict in an evolving global context where conflict no longer has a single logic, and where conflict in the service of values acts against the premises of the state system at the periphery. It is here where the limitations of detente are most obvious. For as the state system has become globalised, so too have the interests of the Superpowers; yet global interests exceed global control, and as the use of force in the service of the state system in the developed world confronts the use of force by states in the service of values in the developing world, a detente based on order, but not change, becomes a victim of the changed context within which it must operate. This dilemma has been shown most clearly with the question of Superpower regional co-operation. *Either* there is no co-operation, and the central balance is seen to become unstable, or else there is co-operation, which in turn leads to regional resentment (not to mention a failure of regional control in a highly unstable local context) which in turn feeds into bilateral hostility and blame at the Superpower level. It is here where detente confronts one of the classic dilemmas of the nuclear age: the relationship between the legitimacy of *power*, and the legitimacy of *values*.

But this inherent instability of detente has been evinced at other levels as well. At the Superpower level, it seemed entirely prudent that attempts should be made to objectify the subjective understanding of the irrationality of war as triggered by the evolving strategic balance and the Cuban Missile Crisis, and to preserve this through measures which would sanctify the perceived strategic stand-off. But this very effort to objectify a subjective understanding has threatened to undo that understanding itself. This flows both from the *fact* of bargaining and the *logic* of bargaining. For by entering into a negotiating context which seeks to compare, balance, and trade various strategic assets, one begins, subjectively at first, and objectively later, to admit, or to feel, that strategic advantage *does* matter – or else why would negotiations designed to arrive at precise measures of balance be necessary? This is compounded by a *logic* of bargaining which assumes its own momentum derived from the adversary context in which it takes place. The assumptions of symmetry, of strategic consensus, of geopolitical similarity of situation (or of indifference to geopolitics!) come up against extant asymmetries, no strategic consensus, and radically different geopolitical situations in a manner which feeds into discussions of unilateral advantage, accusations of betrayal, and assertions of superiority, or the lack thereof in oneself. There begins, once again, to be a re-rationalising of the strategic balance, which, as noted above, acts against the psychological implications of the stable configuration of a mutual *ir*rationality.

But SALT has presented problems at a different level. For by seeking to hive off the strategic relationship from political hostility, two facts became rapidly evident at SALT. First, the semi-autonomous criteria of strategic stability could now be seen to have utility as a *political*, and not merely a *functional*, instrument. If war-avoidance, in its embodiment at SALT, now assumed an independent status, surely that independence could become a political instrument for the amelioration of conflicts at other levels. This, in turn, however implied that a narrow consensus as to war-avoidance – a negative order – could be transformed into a mechanism for positive change. Yet this has manifestly failed, for reasons described in more detail below, both according to the 'internal' criteria of arms control *per se*, and the reciprocal feedback between arms control and its context. As this failure has manifested itself, the hierarchy between SALT and other contexts has evaporated; SALT becomes an issue like any other, rather than an order according to which all else can be related.

This, in turn, raises the fundamental question of 'linkage', in both its latent and explicit forms. (Linkage will *always* exist as a psychological phenomenon even in the absence of contrivance.) For by adopting an instrumental perspective on SALT, one was shifting the criterion of Superpower goodwill from that forum to other contexts, contexts in which the very lack of agreement had led to a desire to disaggregate SALT from political competition in the first place. Clearly, however, linkage is a contextual phenomenon; depending on the issues and balance of influences at stake it may either succeed or fail. Moreover, one person's linkage is another person's dislinkage, and vice versa. For example, if Angola were to be dislinked from SALT, this would imply in Soviet terms, a productive linkage between peripheral behaviour and progress at SALT.

But another fundamental paradox exists with linkage. For such a phenomenon to function successfully, from the American perspective, superiority is an important condition – if not a prerequisite; one could then threaten to opt out of the structure of reassurance and resume the conflict in a mode which would accentuate Soviet weakness. Yet detente had as its premiss the irrelevance of superiority, and indeed parity removed this element of American leverage.

The dilemmas of linkage, however, are exacerbated when the relative positions and attitudes of the Superpowers in the global system are considered. For at the same time linkage has been sought by the United States, the Soviet Union, unlike America, can only see gains in changing the *status quo* in regions where so far it has had few stakes, or no stakes at all; this assumes even greater meaning when one considers that the Soviet Union has never interpreted the correlation of forces as immune from change as a result of strategic stand-off.

But the most illuminating area for elucidating the limitations of detente is centrally, in the nexus of political and strategic problems between East and West, and the Superpowers and their Allies. Here, two particular trends acquire importance. One of the paradoxes of detente has been to disaggregate the strategic confrontation from areas of commitment to central values. In Europe, from the mid-1960s, prior to the development and institutionalisation of the strategic understanding between the Superpowers, one saw attempts to get out from under the framework of security precisely because that security was perceived to allow greater freedom of action. This dialectic was heightened as the Superpowers, bilaterally, sought to reassure each other of their mutual interest in central order. If detente is seen as a commitment to the *status quo* – a negative order – questions immediately arise as to where to *then* move, and if detente is transformed through the mechanism of SALT and linkage, into a dynamic *policy*, as opposed to a legitimated order, questions immediately arise as to first, the different priorities of the Superpowers and their allies as to hierarchies of interest, and secondly, as to the various interpretations of legitimate change as between East and West.

For some Europeans, Germans in particular, detente as *Ostpolitik* sought to derive implications of political and economic co-operation from a premiss of strategic stand-off, to allow, perhaps, the creation of a *Gemeinschaft* of reassurance through a *Gesellschaft* of interaction. But this in turn raises two related issues. What happens to the cohesion of the Western Alliance when different hierarchies of values promote conflict through devising the appropriate linkage to be made between the strategic level and the European level? Should SALT be preserved in order to allow European change, and what if change is not allowed by SALT? Accusations of betrayal, of duopoly, of faint will to resolve Superpower differences at the strategic level, are levelled at the United States by European states and political groups, a context of conflict which is accentuated by the ever-increasing differences in the structure of US-European security, on the one hand, and the structures of economic and political power and interest on the other. (A situation which is itself partially the result of the very success of the security policies which gave rise to detente in the first place.)

However, linked to this are the very different interpretations given to detente by the Superpowers in the context of European relations. For if the United States tends to define security in military terms, and seeks political or societal change once that is attained, and the Soviet Union defines its regional security in ideological and economic terms, there are imposed obvious limitations for the pursuit of detente when defined as political and economic change. Indeed, Soviet policy in Eastern Europe, from the expansion of the cohesion and role of the Warsaw Pact in the mid-1960s through to Czechoslovakia and Poland illustrates the degree to which this is the case. Detente would seem to imply either order – and no change – or change which goes against the central values of one of the participants, which in turn endangers the initial structure of stability. And a structure of order which refuses to allow the consideration of values would seem to be inherently vulnerable to processes of European – indeed global – politics which are imbued with value conflicts of a very potent sort.

Finally, however, the detente process comes up against the tension between Superpower bilateralism and geopolitical asymmetry. For if detente is to be based on the acceptance of strategic stand-off in the context of a threatened irrational retaliation, what happens to Alliance credibility? Nothing, one might argue, as long as that credibility is maintained through that threat. But if, as some Europeans have long argued, deterrence requires more than this – an *empirical* proof of a willingness to go to war – how do MAD, parity, and SALT speak to European needs? Do these phenomena not raise precisely those questions which were subsumed through obfuscation, silence, and institutional palliatives – such as Flexible Response, the Nuclear Planning Group and the vague connection between American nuclear strategies of Assured Destruction and Damage Limitation? It would seem that the very processes of discussion and agreement imply, through objectification, particular *choices*, and those choices are perceived, in European terms, as destabilising. The result is the dissolution of that cohesion of the Alliance upon which the Cold War was stabilised, and the detente was premissed.

In the end, however, detente can be seen to founder at the level of attitudes, rather than realistic and rational assessments of interests and strategies. It is this which indicates the remoteness of detente from social processes of change. For while it could be argued that there are *interests* more in common than conflicting as between Western Europe and the United States, there are countervailing trends in *attitudes* (themselves partly the result of detente and a 'successful' Cold War) which suggests something else. In the end, detente rebounds on the self. For in Europe today, as Pierre Hassner observes, 'one is committed to the more reassuring interpretation of Soviet policies precisely because, deep down, one feels one cannot in fact *afford* to act on the implications of the more worrying one'.[3] Perhaps the 'geopolitics of psychology' is ultimately as relevant to the dissolution of detente as more tangible strategic and political trends.

SALT and the Collapse of Criteria in Strategic Relations

It could be argued that dissecting the process of the dissolution of detente into functional or geographic categories is to repeat the errors of those who saw hope in the construction of the mechanisms of detente in the first place. Indeed, the attempted disaggregation of SALT from the currents of Superpower disagreement at other levels, in order to first stabilise the strategic relationship, and then to use that resulting autonomous structure as a mode of subsequent amelioration of tension – a sort of 'prodigal son' theory of strategic relaxation – has itself been a major source of failure. Thus, arbitrating between the various criteria by which to judge SALT becomes in itself a shadow of the subject under investigation. Should SALT be examined according to the traditional criteria of arms control? Should it, rather, be assessed by less ambitious goals, such as the maintenance of channels of communication and the ratification of pre-existing military capabilities in order to lend some predictability to the strategic system? Or should SALT be evaluated by its contextual impact, according to the impact it has had on overall East-West tensions, on the one hand, and the effect those tensions themselves have had on the SALT process? These questions, however, are best answered by reviewing what, in fact, has transpired.

At bottom SALT is an attempt at objectifying and rendering mutual a particular set of understandings about the nature and requirements of nuclear deterrence. That set of understandings sprang from a particular context in the evolution of American strategic thought; SALT was designed, *inter alia*, to encourage the construction of a set of agreements which would perpetuate that context through the operationalisation of those concepts which the United States had come to see as 'true' and which the Soviet Union would therefore come to accept through strategic maturity and American encouragement. In this sense SALT was both solipsistic and pedagogical: the United States had arrived at a plateau in its strategic thought which was both correct and determining, and the Soviet Union, through the mechanism of negotiation, would come to embrace these revelations.

What were those understandings? In this connection, SALT must be seen as a direct result of the shift in American strategic doctrine away from war-fighting strategies predicated on superiority and rational action, towards doctrine based on the irrational threat of retaliation in the context of a threat to central values. Superiority was deprived of its significance by suggesting that a credible deterrent posture would be based upon a secure relatiatory capability; mutual societal destruction rather than an interchange of military forces deprived marginal technological advantage of any meaning. Both societies therefore had an interest in maintaining the capacity to retaliate and in abjuring from any unilateral strategic effort which was either meaningless, since 'rational' nuclear war would now be met with 'irrational' retaliation (and hence deterred), or dangerous, through the dissolution of credibility of the nuclear threat by depriving either side of its retaliatory capacity through a pre-emptive, disarming strike, or an effective defence.

Both societies were engaged in a clash of interests and values which was fundamental and perhaps irreconcilable; both equally had mutual interests in the avoidance of war through crisis instability, and in creating barriers to the removal of that element of the strategic balance which prevented that value conflict from attaining conflictual form: the capacity to retaliate. The urgency of this task was reinforced by two historical developments: the perceived dangers of the Cuban Missile Crisis which had fed into the doctrinal reformulations which were themselves at the bottom of arms control; and technological change which both suggested the possibility of secure strategic forces (ICBMs but more particularly SLBMs) and the immanence of danger to that capability embodied in ABM technology.

In this sense 'crisis stability' became the central goal of arms control in terms of the intellectual consensus which emerged in the United States in the 1960s. Since a pre-emptive strike was deemed ineffectual in the context of an evolving submarine-based retaliatory force, the only danger seemed to lie in strategic defence. Hence, from the US, and then at SALT, Secretary McNamara attempted to engage in an exercise in strategic evangelism against the ABM evil. In the event, the ABM negotiations proved remarkably easy; this, however, was more likely the result of a feeble and high-priced technology in the first stages of deployment, than any necessary consensus between Soviet and American views on the dangers that technology posed for the deterrence system.

It remained, then, to ask what was left to control, and what principle should guide that attempt. If defensive destabilising technologies were in check, the only alternative remained offensive forces; and if the relevance of superiority was to be denied, then the principle which should guide those negotiations was obviously 'parity'. The United States could scarcely deny the Soviet Union the equal status of strategic Superpower when it had embraced a doctrine which in fact recognised the Soviet Union as a strategic equal, and which simultaneously saw its strategic superiority as insignificant in its fundamental deterrent implications. At the same time all this was going on, moreover, SALT was assuming the crucial role in the detente process which was outlined earlier; it could hardly be abandoned, since to do so would deprive the United States of its entering wedge into wider questions of Superpower relations.

But parity itself is a criterion without strategic meaning. In the context of an assured destruction philosophy it is not necessary for stability, and in the context of crisis stability it is irrelevant; for mirror-imagery of numerical categories does not address qualitative issues which have come to exercise strategic planners – such as Soviet capabilities against fixed ICBMs. Thus, as a result of diplomatic convenience and necessity, parity came to govern the SALT process. As Lawrence Freedman observes, one is left with the proposition that 'a convenient negotiating principle became elevated into a profound strategic insight'.[4]

It is here where many of the destabilising implications of SALT arise, for by enshrining parity one is beginning to say that numbers *do* have strategic significance. At first, this manifested itself in such strictures as the Jackson Amendment which requested that the President 'seek a future treaty that, *inter alia*, would not limit the United States to levels of international strategic forces inferior to the limits provided for the Soviet Union'. This was subsequently embodied in the notion of 'essential equivalence' which directed US SALT negotiators to construct a treaty where there was 'no unilateral advantage to either side'. This, however, was more than a negotiating tactic. Secretary of Defense Schlesinger ascribed a strategic significance to the principle of parity: '[It] is important for symbolic reasons, in large part because the strategic offensive forces have come to be seen by many – however regrettably – as important to the status and stature of a major power'.[5] This, in turn, implied that the doctrinal consensus behind SALT – that numbers did not matter (at the margins) and that there was no strategic significance in superiority – was being eroded by the very process which was seen as bolstering and maintaining that stability, namely SALT.

But this is not all. Parity may well conform to standards of diplomatic practice and jurisprudential equality; it may also sit comfortably with the dynamics of a negotiating context which demands 'objective' standards and the perception of equal concessions. But historically, throughout SALT, both sides have been possessed of asymmetrical forces in the context of rapidly evolving technologies. Since there is no self-evident measure of strategic strength suitable for comparative purposes, partisan bargaining, by concentrating on whatever yardstick accentuates opposing capabilities, generates assessments of advantage which are inherently conflictual. If parity has no objective measure, the field is opened for subjective assertion of unilateral advantage. Moreover, in attempting to create a supportable 'fit' between parity and asymmetrical capabilities, long complicated negotiations have ensued which do violence to the meaning of particular technological trends. Either they are ignored since they cannot be accommodated even within the very elastic conception of parity – or they become accentuated in their importance as each side manoeuvres for negotiating leverage. (An illustration of the former lies in the Soviet advantage in heavy ICBMs; the latter is clearly the case with the Backfire bomber.) The results are agreements where strategic significance and impact are tenuous, or long negotiating processes where suspicion and recrimination impede agreement while strategic reality slips away from negotiating principles.

As parity is accepted as the governing principle of SALT, the purview of SALT expands. For the inevitable partial parities accentuate the significance of imbalances elsewhere, and drive negotiations into areas from which they were excluded precisely because that balance did not exist. The 'intrusion' of SALT into Eurostrategic balances is the most recent example. This partial 'success' coupled with institutional expansionism also allows the direction of military competition into uncontrolled areas, either as bargaining chips (as was the initial intent behind cruise missiles) or as areas of 'fruitful' military endeavour, deemed fruitful often through the very magnification process which accompanies SALT.

But there is a more fundamental problem with parity. By shifting the evaluation of strategic significance away from conceptions of stability based on geopolitical asymmetries – and hence strategic asymmetries – SALT comes up against the differing demands of bilateral deterrence, and deterrence extended to allies who demand empirical proof that the geopolitical asymmetries between Soviet and American positions in Europe are irrelevant to deterrence credibility. This hidden dilemma was postponed in SALT as that process was seen as important by the Europeans for detente purposes, and as the United States continued to exercise a *de facto* superiority, despite the intellectual commitment of SALT to treat such superiority as irrelevant. But as parity was given content through the consequences of American doctrinal revision and the ensuing arms control process, this tension grew. And the very process of negotiating has accentuated this. For by *explicitly* excluding, for example Backfire, from considerations of parity precisely because it was deemed a European (since non-American) weapon, and seeming to do the same with the SS-20, empirical confirmation of European concerns that Superpower sanctuarisation was creeping up upon them seemed at hand. The limitations of parity in the face of geopolitical asymmetries is most obvious at the Eurostrategic level. For if arms control is to continue to revolve around the principle of parity in the context of an overall nuclear balance, strategic reality confronts negotiating convenience. An arms control regime which sees any weapon which can reach the territory of the other Superpower as 'meaningful' – the Soviet definition – is manifestly unacceptable to a Western Alliance whose cohesion has rested on escalation equality, if not dominance. On the other hand, the Soviet Union is equally hostile to a Western conception which would see its Soviet-targeted theatre nuclear force as second-class weapons. The demands of negotiating according to the principle of parity thus come up against the strategic requirements for maintaining Allied cohesion which was the basis for detente in the first place.

There is a disturbing tendency in arms control theory to treat the negotiating process itself as extrinsic to the process of stabilisation. This is not surprising, since that theory sought the deductive application of particular strategic insights; if those insights were valid, then the negotiating process should be both easy, and secondary, to the ultimate effect of 'the agreement' on strategic stability. What has transpired, however, is something diametrically opposed to this hierarchy of considerations: the negotiations *themselves* have done as much to affect strategic stability as have the resulting minimal agreements. In this sense a sort of Heisenberg principle of strategic relations has taken hold; the *very act* of discussing certain issues has altered the nature of those issues in the context of East-West relations generally, and strategic relations in particular. This phenomenon can be seen at various levels, from the role assigned to particular technologies, to the effect on the resurgence of war-fighting nuclear strategies, and the resurrection of heavy issues of Alliance cohesion which had remained dormant precisely because certain questions were not asked.

It is in precisely this manner that the relationship between detente and arms control has transformed itself. From being an attempt to extricate vital matters of mutual strategic importance from the hostility of antagonistic systems of values, SALT has become a victim of that hostility. In seeking an autonomous structure of agreement based on strategic consensus through SALT, arms control came to be seen as a useful instrument in transforming the underlying hostility into reassurance. And in so doing, it became vulnerable to precisely those developments in other contexts which its initiators sought to exclude. SALT became an explicit – or implicit – forum for the consideration of other issues, a context of bargaining between various levels of Superpower hostility. As a direct result of trying to remove politics from the strategic system, politics has been reintroduced. And as the criteria for strategic stability are rendered obscure by technological change, the obfuscation of the original categories of thought (strategic *v*. tactical; convention *v*. nuclear; destabilising technologies *v*. stabilising technologies) and the realisation that doctrines are not only asymmetrical as between the Soviet Union and the United States, but that the United States itself is faced with critical choices between deterrent philosophies (choices which are themselves the result of parity and its adoption at SALT), we no longer know what the criteria for stability are. When this is compounded by the re-politicisation of a technical exercise, *everything* becomes a criterion. As meaning is lost through the dissolution of a consensus as to strategic consequences, that institution which was initially designed to be apart from, but subsequently determining of, Superpower relations becomes a casualty of renewed hostility and instability, themselves the partial result of SALT.

It is to a fundamental re-examination of aspects of this process that this volume is dedicated. By largely eschewing technological minutiae and avoiding compartmentalised debates as to tactical error and short-term policy, it is hoped that a review both of the intellectual universe of those who have responded to – and created – strategic 'reality', and of the basic issues which confront Western strategic planners, will provide some deeper understanding of the strategic present and future. The world of strategy is, almost inspite of itself, in the midst of what Thomas Kuhn would call a 'paradigm shift'. It is by not only answering old questions, but by asking if those questions are really the most important ones to ask, that strategic policy and strategic studies will come to terms with themselves. The contributions assembled in this volume, it is hoped, go some small way to demonstrate the inadequacies of the past, and the tasks for the future.

Notes

1. Robert Hunter, *Security in Europe* (rev. edn) (Elek Books, London, 1972), p. 22.
2. For a cogent analysis of the origins of detente along these lines, see Philip Windsor, *Germany and the Management of Detente* (Chatto and Windus, London, 1971) passim.
3. See below, p. 172.
4. Lawrence Freedman, 'Time For a Reappraisal', *Survival*, Vol. 21, No. 5, Sept/Oct 1979, p. 201.
5. Quoted in ibid., p. 199.

1 ON THE LOGIC OF SECURITY AND ARMS CONTROL IN THE NATO ALLIANCE

Philip Windsor

What is arms control, in terms of the intellectual consensus which had its origins in the strategic culture of the 1960s, particularly in the United States? It is, at bottom, an attempt to create a fundamental and mutual Superpower understanding about the stabilisation of deterrence. This effort at strategic consensus-building is not simply a pedagogical exercise directed towards a shared comprehension of the implications for conflict of the relative configurations of weapons systems on either side. Rather it is an attempt to derive from those technical relationships a degree of political consensus about the way in which those weapons are to be continuously used in terms of the maintenance of the peace – in other words, 'crisis stability'; a logic of Superpower relations is constructed from a logic of conflict. For example, it is still the case in Europe that on most days of the week throughout the year, aircraft take off on opposing sides in the early afternoon, proceed towards each other, and then turn north as they meet along the East-West German frontier around the Elbe, returning to their respective airfields. They are signalling not simply that they share an understanding of the technological capacity of these aircraft in times of war, but that they know what *purpose* continuing confrontation serves; it has become, within the terms of our political understanding, almost as ceremonial as the changing of the guard outside Buckingham Palace. In this sense, the political understanding of the configuration of weapons systems is an essential part of talking about arms control. But one of the difficulties which has emerged from the evolution of the practice and theory of arms control is, perhaps, more central to the current impasse: arms control itself has come to be seen as a means of *creating* that political understanding rather than a reflection of an already existing consensus.

This instrumental nature of arms control follows in a certain sense from the logic of deterrence. Strategic thinkers and practitioners have been preoccupied above all else with the construction and continued operation of a secure deterrent; questions concerning the credibility of a deterrent posture and the preferability of a second strike compared to a first strike doctrine and capability have pervaded strategic discussion. Having followed a relatively simple path of logic in answering these questions, one can begin to understand – or assume – that if the other side also arrives at similar conclusions on the basis of similar concerns and calculations, strategic stability creates a semi-autonomous set of criteria of its own, which in turn act to reinforce a set of political relations. And so the attempt to create an arms control condominium from the logical – and psychological – implications of the various conflicts which have transpired between the Superpowers and their allies becomes, in and of itself, a form of political thinking through the mechanism of a mutually reinforcing Superpower relationship.

This is not to suggest that the logic which has come to inform the relationship between arms control thinking and the wider range of political relationships was a product of deliberate intent at its inception. On the contrary, the original functional, technical approach to arms control questions was an approach which was self-consciously designed to be divorced from the political considerations which have since come to encumber it. However, having appropriated for itself a functional status which sought to accentuate and perpetuate a degree of intellectual and political autonomy, it then became clear that this independence could itself *become* an instrument for the creation of political understanding precisely *because* of this independence.

The paradoxical consequences of this structural situation began to emerge in the particular environment of crisis in the relationship between detente and arms control in Europe during 1968. At that time, three types of conflict came to a head. First, there was a conflict about the meaning of detente. The Superpowers seemed to be engaged in an effort to stabilise and strengthen a certain *status quo* – territorial and political – utilising the manifest implications of deterrence stability as a lever to arrive at a wider, supporting consensus at other levels; the Europeans, on the other hand, sought to use that very stability to promote the peaceful alteration of the *status quo*. Secondly, there was also considerable debate about what constituted the essence of strategic stability, reflected in the debate on ballistic missile defence in the United States, and the first steps towards negotiating limits on strategic weaponry. Finally, there were conflicts which brought the first two categories of debate together in another difference of interpretation: whether political change in Europe could be accompanied by strategic stabilisation through agreement between East and West. All three questions seemed to be answered by the Soviet invasion of Czechoslovakia in 1968. That action put paid to the hopes of Europeans that political change could, indeed, accompany strategic stability. On the contrary, the Russian tanks argued that political change which threatened the self-defined central values of one Superpower could not be married with the effort to secure strategic stability; the latter could only be maintained by ensuring the maintenance of political stability, which in turn meant no change.

The reaction of the United States to this situation was fundamentally a drawn-out, complicated, soul-searching affair, in which the Administration initially postponed the SALT talks, the initiation of which it had already agreed to. (If one senses a certain similarity here to what occurred in relation to the ratification of the SALT II treaty, this is more coincidental than real.) The delay in the commencement of the SALT I talks was accompanied by a recognition that strategic arms control could no longer be seen only as functional in terms of the stabilisation of the deterrence structure; they would also be inevitably functional in terms of the creation of a new political relationship. Thus the SALT process, initially perceived as a means of getting *away* from politics, paradoxically became central to the construction and management of an emerging Superpower detente. Indeed, Secretary of Defense McNamara directed the first American negotiators at SALT to spend their early sessions discussing not hardware problems or issues of arms race and technical stability, but rather the philosophy of deterrence and the building of a common universe of conceptual, almost metaphysical, understanding which could *then* serve as the means of approaching both more narrow technical questions and questions of a political nature.

In Europe, the effect of the invasion of Czechoslovakia was equally profound. The early pattern of detente on the continent – epitomised by the approach of President de Gaulle – had been characterised by a strategy whereby the European states were able and willing to exploit a relative degree of stability in Superpower relations by trying to encourage a greater degree of freedom of action in Eastern Europe. This gave way, in 1968, to a type of approach more easily identified with the name of Willy Brandt than President de Gaulle. It was, in essence, like a prolonged and skilful exposition of the art of Ju-Jitsu: the great weight of the Soviet Union was thrown on its back by the smaller power, who made it clear that if the USSR was serious in its desire to negotiate successfully with the Federal Republic of Germany over certain points of legitimacy and geography – such as post-war borders and the status of Berlin – and improve its economic ties with the West, then the interests of Bonn in creating a 'dynamic *status quo*' in East Germany, and Eastern Europe more generally, would have to be recognised. Brandt's dynamic *status quo* was one in which the issues of security would be compartmentalised from those of political and economic interchange; the former would remain relatively static while the latter would be legitimate ground for movement and change over an indeterminate period of time. Thus, the detente of Willy Brandt was a beneficiary of the invasion of Czechoslovakia as were the SALT negotiations, through the medium of American reaction and reassessment.

When the SALT negotiations began in 1969, there emerged a vision of the nature and purpose of arms control which did not derive from its functional base, and where that functional base was transfigured into the form of an attempt at generating political understanding. The narrower instrumentalities of arms control became the vicarious vehicles for dealing with a whole range of broader European – and global – issues. Throughout the period during which the SALT I treaty was negotiated, the United States became increasingly self-conscious of its use of SALT for broader purposes; President Nixon's State of the World Message to Congress, for example, explicitly advised the Soviet Union of the connections which would be drawn between Soviet (non-) accommodation in the Middle East, and progress at SALT. SALT and other arms control fora became a holding operation whose major purpose was to serve as a guarantor of a particular strategic order, on the one hand, and an institution whose value to both Superpowers would allow changes, on the other hand. This relationship between order and change became a fluid and self-transforming dynamic which was worked out through the SALT process, as both states utilised these negotiations and their linkage to other issues as a tool for the establishment of what fell under the rubric of essential 'order', and what lay in the domain of permissible 'change'.

SALT, in other words, became increasingly political, and for a time it seemed to succeed – for both 'good' and 'bad' reasons. The United States could persevere in its intended mandate for SALT so long as it was strategically superior in most categories; Soviet inferiority created incentives for the apparent fulfilment of American negotiating objectives in, for example, the Middle East, where a clear US advantage in MIRV technology may be seen as having allowed a successful linkage of SALT with issues of regional security and Superpower behaviour. However, the continued achievement of American aims was predicated on more basic – and questionable – presmisses: the understanding that if the Soviet Union was engaged in the vast political dialogue of which the *United States* understood that SALT was the centrepiece, then the Soviet Union would accept that SALT was the centrepiece as well. This, however, might hold only so long as the US was possessed of a very clear edge of strategic superiority.

But that superiority has, of course, disappeared, or at least become less a canon than a point of strategic debate. Comparatively simple questions of weapons configurations are turning into questions of different categories, in which the inherent tensions of deterrence are posing new questions of credibility and stability. Being superior in one category has come to mean inferiority in another – as with theatre nuclear forces – while the existence of stability in one functional or geographical area can mean creating the risk of instability in another – as with inter-continental parity and peripheral conflict. In consequence, the attempt to preserve – or indeed create – a balance between order and change has become increasingly difficult to maintain, just as the secular trends which must be managed have themselves made that task more onerous. This process of Superpower strategic overload began in the early 1970s, perhaps during the 1973 Middle East war, and while it took some time for objective change, in Marxist terms, to become subjective change, the world has been living with its consequences ever since.

Prior to examining the implications of this breakdown, however, it might be useful to briefly examine some of the criteria by which this process can be seen to have occurred. It was a basic premiss of classical strategic and arms control thinking that the nature of strategic stability was easily ascertained and that, indeed, this constituted an 'objective' element of the nuclear universe to which all else could be related. It only remained, therefore, to 'educate' both sides as to the truth of the matter; any difficulties to be encountered were likely to arise primarily through the negotiating process – the means – rather than the end itself. Thus parity was seen to be an empirical criteria by which to judge the existence, or lack thereof, of stability, and the American SALT negotiators argued for the institutionalisation of this concept through the arms control process. This argument was conducted through an exquisite, increasingly etiolated, ptolemaic logic in which epicycle after epicycle of technological change was brought into a kind of cosmos of overall strategic stability, despite the fact that there was no parity in any conceivable category whatsoever. This relationship between parity and stability, however, was defended not simply in terms of what one side could in fact 'do' to the other in purely technological terms, but more importantly in terms of the perceptions of political tensions which arose from relating parity to stability.

But the relationship between the criterion – parity – and the value

– stability – is more complicated than this. For example, at the early stages, one of the prime instruments of strategic instability was very clearly ABM technology. Thus, the fear that ABMs could encourage a first-strike posture helped lead to a quick SALT I accord. But are ABMs destabilising when, due to *initial* backwardness of technology, a capability is then developed to destroy a large percentage of the other side's land-based missile force as a consequence of subsequent technological advances? Might it not be argued that the same criteria which argued in favour of ABM limitation at a particular state in relative technological development now argue in favour of the possession of a limited ABM capability by one side to remedy the disadvantageous – and destabilising – changes in relative weapons quality and quantity which have since transpired. In this sense, the relationship between parity and stability, and what, indeed, constitute the proper instruments of stability are constantly changing, doing violence to the static conceptions which have been enshrined in arms control dogma and institutions.

These developments might have been manageable if there existed a stable political context within which to discuss such questions. But this context has itself become less stable precisely because of the success of the initial assumptions of SALT. SALT did create a degree of agreement on strategic stand-off between the two Superpowers, which in turn acted to preclude at least some measure of the risk of going to nuclear war. It also, thereby, increased two kinds of potential danger. First, in peripheral areas, as the Soviet Union began to redress its technological position *vis-à-vis* the United States at the strategic level, it began to feel freer to act at lower levels of violence and commitments to values than those which comported the risk of nuclear war. About four years ago, I was asked by a Soviet Colonel-General what I thought had been the most important change in international politics over the previous few years. Failing to supply the correct answer to this question, the visitor said, 'No, the most important change that has taken place is that we are no longer afraid of you.' He was articulating a qualitative change in the psychological response of the Soviet Union to its environment; part of that response was the recognition that strategic parity was being secured, and that Soviet advantages in other areas at different levels could now be exploited with greater ease, flexibility and freedom than had been the case in the past. Thus the first product – and problem – of a successful SALT was that by creating at least the promise of strategic stability, it simultaneously created the danger of the exploitation of instability at lesser levels, and thereby the multiplication of global conflicts in contexts which until that moment had had very little to do with arms control. The Middle East, Africa and South-West Asia are all now more open to greater instability through the very stability initially created by SALT.

The second category of danger follows from the first. To what extent can the maintenance of a central political relationship, which in turn reinforces strategic stability, be a priority for two states which are engaged in serious competitions over a series of other questions? Both the USSR and USA, preoccupied with direct and indirect contests over political influence, natural resources, and ideological legitimacy, are locked in an overall confrontation which engages the very nature of what it is to be a Superpower. Moreover, the fundamental asymmetry in attributes of the two states may encourage the Soviet Union – technologically weak, economically in decline, and ideologically on the defensive – to compensate for such forms of weakness in that dimension of strength which it posesses in full measure: military power. Furthermore, that military power is increasingly capable of successful operation at lower levels of violence than those which involve the risk of nuclear war. In this sense, such a Superpower may be tempted to exploit change and disturbance in the world which will not threaten SALT because the other Superpower insists that they *should* not threaten SALT.

So, during the SALT II negotiations, increasing trends in favour of political change and instability converged with the very period when the United States was endeavouring with accelerating vigour to secure an agreement. Gradually, the SALT process itself became hostage to the political process which it was originally designed to create. In consequence, questions multiplied in Washington about whether the United States should proceed with SALT in a period where the context of political change and the very nature of the Superpower relationship was making for increasing instability. So the whole connection between SALT and the broader process of international relations has turned turtle. Originally, other political relations were hostage to SALT; SALT has now become hostage to other political relations. Compounding these developments, all this has been going on precisely when the criteria of strategic stability are becoming progressively harder to identify and objectify through international agreement. At the time the SALT II treaty was reached, it was generally accepted that it was practically irrelevant to any of the significant technological changes occurring in the strategic competition between the two Superpowers. The stability of SALT itself has thereby fallen prey to technological flux in a manner which yields to a variable form of unstable dependence on the nature of the political relations between Moscow and Washington.

It is with this context in mind that recent developments in European security must be evaluated. Helmut Schmidt, in his 1977 Alastair Buchan Memorial Lecture at the IISS, suggested – perhaps with a touch of hyperbole – that the existence and imminence of SALT agreements could make the question of grey-area weapons and confrontations in Europe increasingly problematic. To the extent that strategic stand-off is successfully negotiated and embodied in international agreement, all the old questions of strategic credibility and escalation dominance would acquire a heightened urgency.

West Europe has never developed a methodology for dealing with questions arising from the military relationship between the two halves of the continent. This has its origins in the 1950s, when the West European abjured from the sacrifices required to invest the money and manpower necessary to mount a credible conventional defence; it was clearly preferable to whistle up the inexpensive alternative of American nuclear and air power. At that time, these states congratulated themselves for the effectiveness with which they had tied the United States to the defence of Europe. The result today is that as Washington is confronted with more and more insurmountable difficulties in the defence of Europe, partly through strategic parity and partly through Soviet theatre superiority, so the Western powers have found themselves virtually incapable – spiritually and materially – of creating any form of Western defence policy whatsoever.

The concept of 'credibility' is still a key criterion because Europeans, in characteristically logical manner, rightly wonder what America would be able or prepared to risk doing on behalf of Western Europe. But the European defence dilemma has been intensified by the posture adopted by NATO for the defence of Germany between 1955 and 1957. It could well be argued that a strategic position which did not run East-West but ran North-South would give the Europeans more room for tactical manoeuvres, more warning time in which to deal with the threat and possibility of a Soviet or Warsaw Pact attack, and greater flexibility through the multiplication of options, but this possibility was eschewed for political reasons, and because of the manner in which Germany joined NATO. The result is the perpetual raising of that fundamental yet foolish question, 'When do we go nuclear, and how far does the enemy have to go before that decision is made?' This question is obviously unanswerable. It was also fairly unimportant so long as the

Soviet Union was in a strategic position inferior to that of the United States, or, in conditions of parity, was able to agree on the political conditions required for the maintenance of stability. But in conditions where these political conditions are ill-defined and where parity has melted into superiority, the questions of Soviet strategic advantage become acute, as does the issue of what Moscow might be tempted to do with its increasing power in Europe over the next few years during what Kissinger has labelled the window of opportunity. But this period of opportunity coincides with one in which one no longer knows what the criteria of political or strategic relationships are.

If the Western powers, in these conditions, have no adequate riposte to Soviet strategic potential, they will be inclined to opt for greater political intercourse with the Soviet Union, almost, in effect, buying them off. This should not necessarily be seen as conforming to the squalid practice of paying *Danegeld* although this has certainly been an aspect of European attitudes and policies at times. Rather, this should be seen as an attempt to wean the Soviet Union from its apparent current political intentions through the creation of a network of forms of interdependence such that it occurs to Moscow that it is not worth their while to attack or pressure Western Europe, and that every effort should be made to encourage, or at least preserve, West European autonomy and prosperity and enter into a co-operative relationship involving the whole continent. It is unfortunate and unfair that this West European attitude, which one could sum up as the search for an organic relationship, is precisely the perspective which the European press criticised so fiercely when it was articulated by Helmut Sonnenfeldt in his remarks about Eastern Europe. The two approaches display similar contours, however: an attempt to create a relationship between Eastern Europe and the Soviet Union, and between Eastern and Western Europe which would minimise the importance of the security confrontation and maximise the importance of other forms of political, economic, technological and cultural intercourse. The high point of this strategy was attained through the success of the 'Brandt detente' whereby the *Ostpolitik* became an attempt to legitimise change in Europe in such a manner that it did not trigger Soviet perceptions of threat to its vital security interests. But this strategy meant, even when it succeeded – and it certainly did not always succeed in terms of Brandt's original expectations – that such hopes of transformation through rapprochement can rapidly be thwarted by the dour and unyielding East German policy of *Abgrenzung*. In such conditions, it quickly becomes apparent that there is a greater necessity to rely on the maintenance of a set of political relationships with the Soviet Union.

But in these circumstances, what *then* is the importance of arms control in Europe? Is arms control an instrument of that relationship or is it an outcome? Should an attempt be made to confront the Soviet Union with a choice between an acceptance of Western ideas on MBFR and the proper relationship between Superpower strategic forces and European forces on the one hand, and the severance of a relationship between politics and arms control? Certainly, the history of MBFR does not speak well for the influence of arms negotiations on political relations on the continent.

Perhaps an effort should be made to distinguish between such matters and to proceed from asking what kinds of concessions we can make to asking what kind of concession *they* can make. In this connection, it is possible to imagine the Western powers suggesting certain economic concessions, with the Warsaw Pact offering certain security concessions. Theoretically, this approach has considerable plausibility, but it is extremely difficult to operationalise. What occurs in economic terms usually has its greatest effect over the long term; what takes place in security terms usually has to be shown to be viable over the short term, given the interests and sensitivities at stake. Moreover, how can such a strategy be organised among a set of social democracies where there is an accepted disaggregation of the pursuit of economic interest and gain from issues of security? It is difficult enough to dictate to Olympic athletes; problems multiply when dealing with more powerful, competing, private firms over a long period of time.

While it may be possible to envisage or measure a viable trade-off at the European level between security and other issues, difficulties are likely to multiply when the Superpowers are brought into the process. Questions of this order arose during the CSCE negotiations. The Federal Republic of Germany was at that stage anxious to create a relationship (not a linkage or *Junctum* as the Germans would term it, but a relationship – a *sachlicher Zusammenhang*) between questions of economic, commercial and political relations, on the one hand, and security relations in Europe, on the other, such that the two sets could be discussed side by side. But this was seen by the United States, precisely in terms of the strategic management which SALT implied, as a threat to the priority of those talks. Gromyko, after all, had said so, and had told Henry Kissinger in Vienna that the CSCE was hampering the successful conclusion of the SALT II agreement. Thus the West European effort to create a relationship between political and economic concessions and security concessions was thwarted through an unavoidable difference in hierarchies of values between the Superpowers and their European allies. In practical terms, there was very little chance of weaving a relationship between arms control and the wider context of political consensus. Yet the SALT talks were *still* seen as the primary mediator between order and change in the international system, a view to which the classical theory of arms control had led.

In consequence, by the end of the 1970s, many different expectations had been explored and exploded: the hope that SALT could widen its purview and effect from functional arms control to a political relationship; the hope that this relationship could be cemented by overt linkage between progress in SALT and events and behaviour in other areas of the world; the hope that such connections could lead to a form of control of the configuration of arms in Europe; and the hope that the arms configuration in Europe could itself be directed by political and economic instrumentalities. And *all* these hopes were dashed well before the Soviet invasion of Afghanistan.

It was clear by the end of 1979 that the American appreciation of Soviet intentions and behaviour had undergone a long process of transformation which rendered the Afghanistan question more a confirmatory event than a catalyst for a paradigm shift in American attitudes. New forms of containment policy had been actively considered in other areas of the world a long time prior to the invasion. (It could also be argued that the Soviet Union was, in a sense dragged into Afghanistan, and that the invasion did not betoken a fundamental change in Soviet geopolitical intentions, but rather conformed to extant conceptions of Soviet interests and security.) What Afghanistan *did* betoken, however, was a fundamental realisation of the way the geopolitical discord and mutual (mis)perceptions of the Superpowers were now to be worked through. In consequence, the logic of arms control in Europe virtually disappeared.

The West Europeans have found themselves involved in a tremendous crisis of confidence within the Alliance. If one examines the condition of NATO in December 1979, things were going surprisingly well. There were still many problems: the window of opportunity was on everybody's mind and questions of the reaction to that asymmetry in the strategic balance posed the dilemma as to whether the modernisation of theatre nuclear forces would in fact delay or destroy the possibility of serious arms control negotiations, or whether the threat to develop and deploy modernised theatre nuclear forces might not, in fact, speed them up. But the decision in favour of modernisation, with a connected commitment to seek arms control agreement, was arrived at with surprising co-operation between Germany and the United States considering the depth of antagonisms between them. The United States was grateful to German virtuosity in pulling the Alliance together over the LRTNF question, and many countries were beginning to take real action to fulfil the commitments inherent in the Long Term Defence Programme. It was difficult to predict that within a matter of weeks the Alliance would be facing the most fundamental crisis of its entire existence.

That crisis was based on a qualitative change in the terms in which the politics of the Alliance had, until then, been broached. In the past, the politics of NATO had been based on the question of how to establish the determinants of American credibility itself. Debate, worry, and conflict revolved around the search for the criteria by which Europe could judge American constancy: hence the history of issues ranging from the McNamara Doctrine to the implications of SALT, grey-area problems, etc. But these conflicts have been dealt with with greater or lesser degrees of success through empirical solutions ranging from the Multilateral Force to the Nuclear Planning Group and the usefully vague doctrine of flexible response. At the end of 1979, the Alliance had entered a more profound crisis of confidence for two reasons. First, it was the principle of credibility itself which was at stake, and secondly, it was no longer a question of *American* credibility; it was now a question of *European* credibility. No longer was the Alliance preoccupied with its traditional chronic discussions of whether and in what manner the United States would come to Europe's aid; America now began to wonder if Europe was worth defending. It is because of this basic transformation of the locus of the burden of psychological proof which indicates that the Alliance is now in a period of greater structural crisis than in any of its previous dilemmas like Suez, and the 1973 Middle East war, and every other Middle East episode which has always split the Alliance much more than anything which has happened in Europe.

These questions of credibility now mean that the difference between how to approach arms control questions in Europe and how to approach arms control questions at the Superpower level has become far more acute. The Europeans, rightly and unavoidably concerned primarily with arms control *in Europe*, must focus on, accommodate, and build upon, the inextricable link between European political relations and arms control which past history has shown to be insurmountable. But this attempt to create an arms control approach to wider

European issues is in direct conflict with the equally right and unavoidable American need to create a containment policy which is first of all strategic, and which might allow a return to arms control in the original functional sense of the term. President Carter, for example, always said, even when proclaiming the Carter Doctrine, that it was essential to continue with arms control negotiations with the USSR. At the same time, however, he suspended the SALT ratification process, no doubt partly because he realised that SALT would probably not get through intact, but also because the continuation of that process would then be seen as political legitimation of Soviet behaviour since SALT, by that time, had become part of the political currency of exchange between the two sides. To suspend SALT and then to insist on the vital necessity of getting back to the arms control negotiating table was, it would appear, an attempt on the part of the United States to say that arms control must take its *functional* place once again, even if – perhaps because – it had fallen short of the political functions which were attached to it before. (To answer this failure with reinvigorated linkage, as seems the initial intent of the Reagan Administration, is to speak against the testimony of recent history.)

This is perfectly logical in the American context. In the European context, however, this approach could provoke once again all the old questions, which were so dominant in European politics for so many years, of how to approach questions of arms control through political relations with the USSR and the Warsaw Pact. Yet there is no emerging logic by which one can say that there are certain sets of arms control proposals which could create stability. The changing criteria of strategic stability have now become caught up in a changing political context of the utmost potential violence. It is therefore no longer possible to declare, as one could in the past, that it is possible to create *either* political stability or indeed strategic stability through the instrumentality of arms control.

Arms control in Europe must explicitly conform to its true role – as an overtly political act. Arms control between the Superpowers must tread where it is necessary, and where it is feasible – as an overtly functional act. These two sets of values and contexts can no longer be married in the way that previous arms control practitioners tried to marry them through the relationship between MBFR, SALT and CSCE. Questions of categorisation, technical linkage, and institutional co-ordination will continue to interact in a functional manner. But these questions should be seen as second-order issues. In the meantime, the political nature of arms control in Europe and the functional criteria of Superpower arms control should be recognised, and the acceptance of that distinction treated as a matter of primacy.

2 ARMS CONTROL: THE POSSIBILITY OF A SECOND COMING[1]

Lawrence Freedman

Not long ago arms control appeared as a wholesome thing. Everyone, even hawks and weapons designers, would insist that everything they did in the military sphere had arms control objectives in view. It was a selling point: new weapons were promoted by reference to the virtues of stability and verifiability, or else by the real hardliners as a 'bargaining chip' for future negotiations; future strategic moves were always conditional on the progress of negotiations; doctrinal pronouncements acknowledged the possibility of lasting peace, as if embarrassed to dwell too long on the nasty business of war-fighting.

Now all this has changed. The few unrepentant arms controllers appear as the Keynesians of strategic studies, accused of misguided idealism, political naïvety and wilful meddling with national security. In the dock some plead guilty and promise that they are now reformed characters, properly suspicious of the Russians and enthused over M-X, while others merely look crestfallen wondering where it all went wrong.

Many who had placed their hopes for a more peaceful world on arms control now consider it to be a cynical manoeuvre, designed to deflect energies away from real disarmament. The arms control decade failed to dent seriously any military programmes. Such achievements as have been managed are little compared with the actual record of a deterioration in East-West relations and the emergence of dangerous new technologies and philosophies to match.

Everyone seems to agree that the effort has failed, pointing to the refrigeration of SALT, the tedium of MBFR, the shrinking comprehensive test ban, the abandoned efforts to regulate conventional arms transfers or the deployment of forces in the Indian Ocean. Intellectually and politically, arms control is exhausted. If it is not yet diplomatically exhausted that is only because diplomats appear inexhaustable. To many the real arms control problem is now how to retire gracefully from the whole sorry enterprise and concentrate on the 'real business' of defence or disarmament (depending on the point of view).

The prevailing liberal mood when an arms control agenda was first identified in the late 1950s and early 1960s showed a preference for sophisticated management of the *status quo* rather than grand designs for restructuring national and international society. This was the age of the 'end of ideology' and the distrust of the radical and the passionate. The cool, the pragmatic and the technical were found more congenial. The terminology and prescriptions of the arms controllers were pitched just right: a minimum of moralising; a modicum of technocratic jargon; a dash of realpolitik; a pronounced streak of practicality and a prospect of careful but visible progress.[2]

Their political strength derived from appropriating disarmers' objectives (peace and co-operation) while deflecting military opposition by showing slight interest in actual disarmament. Over time this compromise became difficult to sustain. The disarmers grew frustrated and came to see arms control as a Superpower confidence trick, designed to avoid genuine change, rather than simply a moderate version of their own programme.[3] Meanwhile the military became irritated with any interference in force planning, however marginal it turned out to be in practice. The military establishments of both sides came to suspect that arms control was proving to be far more of an imposition on their plans than on those of the adversary.

The desertion of both hawks and doves might have been averted had arms control maintained its early momentum. But the spate of negotiations set in motion in the late 1960s and early 1970s became bogged down. Few agreements were reached and, on the whole, they represented the minimum negotiable. The glacial movement encouraged cynicism. Publicity given to deviations from the original, optimum Western positions in the various discussions allowed those alarmed by these deviations time to generate opposition.[4] With growing scepticism over the sincerity of the Soviet interest in detente, this opposition found fertile ground, particularly, but not solely, in the United States.

In public policy in general there has been a decline in the influence of liberal sentimentality and a preference for hard-headed conservatism. The doubts about the economic benefit of government intervention in the market are matched by those who believe that attempts to regulate strategic affairs will end in distortion rather than improvement of national security. This sceptical mood, combined with a readiness to contemplate a long haul in East-West relations has created an atmosphere hardly conducive to arms control. The consensus seems so overwhelming, that the necessary first question must be 'Has arms control really failed so badly?' To answer this we must look at the various objectives associated with it over the years.

Strategic Stability

The essence of arms control theory is the notion of incomplete antagonism. Unlike disarmers, arms controllers did not argue that the cold war was all a terrible mistake based on the fear engendered by the arms race. East-West conflict was recognised to be based on a fundamental, and possibly irreconcilable, clash of interests and values. However, this antagonism was incomplete. There were mutual interests, of which the most critical was in avoiding a general nuclear war.

Much followed once this mutual interest had been recognised. The Superpowers had to work out ways to confront each other without letting matters get out of hand. This involved:

(a) Learning to accept limited political objectives in any particular crisis, and not pushing matters to a decisive showdown;[5]
(b) Developing conventions of crisis management that might allow for diplomatic compromise;[6]
(c) Configuring military forces in such a way that their use, if necessary, would be deliberate, considered and contained.

It was assumed that much of this would develop tacitly, without prompting, through common sense and experience. With (a) and (b) this has largely been the case. It was also assumed that the development of these implicit conventions, guidelines and understandings would be quite consistent with preparations for conflict.

The basis of arms control is found in (c). In a crisis or the early stages of a war this meant avoiding the loss of central control over actions on the ground lest excessive violence flare up and spread because of some hot-blooded commander or an unruly client-state. However, a more critical requirement was to encourage rational strategies for the use of armed force in which there would be no military profit in actually initiating war. The outbreak of the First World War, in which the need to mobilise first led to the premature use of force, was taken as a salutary lesson. No new general war must be allowed to start before all diplomatic options had been exhausted. It had to be possible to hold back on military action without fear of being defeated in a surprise attack.

This concern developed in the context of a growing fear that developments in nuclear weapons' technology were putting a premium on the first strike. The only possible source of 'victory' in nuclear war would be destruction of the adversary's nuclear capabilities in a disarming, pre-emptive strike and/or catching any retaliatory force, once launched, by effective active defences. Schelling and others wrote of a 'reciprocal fear of surprise attack' in which a crisis could get out of control by each side worrying that if he waited to employ his nuclear arsenal it could be caught by enemy pre-emption. In the United States, memories of Pearl Harbor encouraged this preoccupation with surprise attack.

All this led arms controllers to argue for measures such as 'fail-safe' control systems, protected command and control facilities and 'hot-lines'. Second-strike systems, that is those that could neither threaten nor be threatened by the adversary's strategic forces, were to be encouraged. The exemplary second-strike weapon was the ballistic-missile carrying submarine (SSBN). Stability was seen to stem from mutual vulnerability.

Arms controllers assumed that the nuclear balance had created a stability that, while uncomfortable, could well be durable. The objective was to reinforce the *status quo*. There was little room for disarmament in this scheme of things – indeed too much of it might undermine stability by reducing the destructiveness of war to tolerable levels.

The intention was to facilitate the management of the balance of terror. This did not necessarily depend on negotiation but could be implemented through national policies. With some exceptions such as the 'hot-line' agreement to establish an emergency communications link between Moscow and Washington, many of these measures were adopted unilaterally, at least in the American case. When Robert McNamara announced plans to improve command and control mechanisms so as to reduce the risks of accidental launches of nuclear weapons he did it in such a way as to encourage the Russians to follow suit.

Because of the self-evident value of protecting at least one part of the deterrent from surprise attack little encouragement was needed to build up the SSBN force. Both sides (though the USSR lagged behind) did this, and in so doing, established the sort of strategic relationship required by arms control theory. With a secure means of retaliation there could be no military premium in launching a surprise attack.

This relationship, which came to be known as one of Mutual Assured Destruction, was further reinforced by the difficulties experienced in developing ballistic missile defences. In the 1960s both Superpowers were embarked on defensive programmes which, if successful, might have had a disturbing impact on the strategic balance by making destruction less assured. On the other hand, failure would be an expensive waste of effort. As it turned out the major advances in radar, computing and interception over the 1960s were trumped by advances in offensive capabilities, particularly MIRVing. The conditions were therefore ripe for a negotiated agreement. This was achieved in 1972 in an ABM Treaty which is now considered the purest achievement of modern arms control.

Once ABMs were under control the strategic balance was rendered reasonably stable – certainly in terms of the theories of the 1960s. There was not very much left to be controlled. The only major development which might still require control – MIRVs – escaped effective control. The potential problem they posed was in making it possible for one ICBM force to eliminate another ICBM force of a broadly similar size, with a large number of warheads left over to spare. MIRVs were not perceived as a 'problem' until they were virtually ready for deployment in the Minuteman III ICBM and Poseidon SLBM by the US,[7] but then the majority view in Washington was that they were a unilateral American advantage and not to be conceded short of some spectacular concessions from the other side. Anxiety only really developed when the Soviet Union developed its own MIRVs.[8] Then the Soviet advantage in both the quantity and throw-weight of their ICBMs meant that they were able to fit large numbers of accurate warheads, and so gain the full advantage of the counterforce potential of MIRVing before the United States. Unfortunately, serious constraints have proved elusive. The various attempts to inhibit Soviet MIRV development – the sub-category of 'heavy' ICBMs in SALT I, the ceiling for MIRVed delivery vehicles at Vladivostok, the ceiling for MIRVed ICBMs and the fractionation provision of SALT II – all failed to prevent the emergence of the ICBM vulnerability problem.

As a problem of strategic stability this has certainly been exaggerated in that the grounds for believing that the capacity to destroy ICBMs without something comparable for SLBMs creates a major advantage are somewhat flimsy. The arguments surrounding whether SALT could have done more to alleviate the MIRVing problem may have endowed the ICBM vulnerability issue with greater significance than it deserved. It is also an issue that is extremely difficult to cope with, by military measures alone, without arms control.[9] Nevertheless, the proliferation of highly accurate warheads has been the most destabilising development of the last decade and arms controllers have been able to do little about it. At the same time it is arguable that the most stabilising developments – the strength of SSBNs and the weakness of ABMs – have come about as a result of technological and military logic rather than because of any deliberate arms control effort.

Certainly strategic activity not touched by international agreement is not wholly 'out of control', driven by primitive passions of militarism and nationalism. Nuclear war has patently been avoided and, while the nuclear arsenals have undergone substantial modernisation since 1960 this has been, at least in the West, less burdensome economically than in the 1950s, and directed towards greater precision in attack (and thus potentially less indiscriminate destruction). Much of this happened quite independently of formal arms control. Furthermore, the critique of the whole concept of strategic stability, and mutually assured destruction, has been largely based on the view that with lots of precision warheads opportunities exist for more sophisticated forms of nuclear war-fighting. It is argued that this offers advantages without resort to all-out exchanges, and that the USSR was better prepared for this by virtue of both its political philosophy and military doctrine. Without entering into the argument on this matter,[10] we can note that it hints at something of a paradox. If the balance is unstable then arms control needs to be applied with some urgency, *but* if the causes of instability are to be found in Soviet attitudes to war as much as particular capabilities then the scope for successful arms control is limited. Those who believe in a fragile balance do not tend to be those who see any value in seeking accommodations with the USSR. If, on the other hand, the balance is essentially stable then arms control, as traditionally conceived, is not vital. Either way, there is little place for arms control.

Parity

The formal negotiations of the 1970s have been based less on pure arms control theory than on a judgement on the political effects of the more visible aspects of the strategic balance. The presumption is that whatever the strict military relevance of a particular disparity in capabilities between two sides, once it has attracted notice it gains a strategic significance by affecting the credibility of and confidence in a deterrent posture. The proposals have tended towards codifying 'parity', that is creating a visible equality in force levels.

This is an appealing notion for diplomats because it invokes the principles of equity and fairness that are useful components of any international agreement, and because it can be based on levels and types of hardware that, in so far as they can be easily identified and counted, allow for diplomatic trading and measured agreements. 'Parity' has a precedent in the inter-war disarmament efforts, particularly the Washington Naval Conference of 1925.

The major difficulty with this approach has been that it has not been possible to negotiate balance out of imbalance. In negotiations it is equality in concessions that matters rather than equality in outcome. It has thus far only been possible to ratify a pre-existing parity. Furthermore, the most effective negotiating formula has been one that required minimal changes in the force plans of either side. Negotiators have sought for yardsticks of military strength which yielded a suitable compromise without being wholly devoid of strategic content.

During the 1970s the weapon systems at the centre of SALT – ICBMs, SLBMs and bombers – were both suitable for counting and possessed in broadly similar quantities on either side. This made successful trading possible, but not without difficulties. The agreements on offensive arms had to be forged against a background of a rapidly advancing technology and growing variations between the force structures of both sides. Thus, though SALT agreements were reached in 1972 and 1979 (and partly in 1974) they could be criticised for controlling very little and, then, only the technology of the past rather than that of the future, and also for achieving formal agreement at the expense of substantive value. Moreover, the negotiations themselves were long and difficult. The asymmetries were sufficient to make the trade complicated – to the extent that obscure details suddenly were magnified into matters of central importance in East-West relations.

Once an attempt is made to establish parity in one area of arms competition between the Superpowers then the process becomes difficult to stop. Without comprehensive parity, military activity can always be redirected out of the area covered by partial parity. Arms control has come to be expansionist, embracing more and more types of weapons. SALT I was conceived with reference to ABMs; in its first phase it included ICBMs and SLBMs; by 1974 bombers were involved; by 1976, cruise missiles had arrived and systems based in the European theatre were implicated. In 1979, NATO proposed further negotiations on some sort of parity in long-range theatre nuclear systems and preliminary discussions began in October 1980. In doing so, NATO attempted to exclude aircraft, but the USSR is insisting upon this. It is hard to see how the logic of introducing short-range and non-American systems, once the 'parity' game is to be played, can be avoided.

The proposition that 'negotiability' requires a pre-existing parity indicates the source of the problems faced in Vienna at the talks on Mutual Force Reductions (MFR).[11] The talks began in 1973 in conditions of patent asymmetry. Not only did the Warsaw Pact enjoy superiority in numbers of tanks and troops, but also it gained enormous advantages as a result of geography. Warsaw Pact reserves would always be close at hand and enjoy ease of access because of the proximity of the USSR. US reserves would have to travel great distances by sea. Furthermore, there were limits to the troop reductions that either side could contemplate, determined more by consideration of alliance cohesion than the 'threat'. The USSR bases troops in Eastern Europe in part to watch for signs of wavering loyalty in the socialist camp. NATO is unable to contemplate either the neutralisation of West Germany nor the withdrawal of allied troops from its soil below the point where there ceases to be a demonstration of alliance solidarity.

In seeking parity, MFR negotiators have had to avoid tackling the geographical problems, for there can be no compensation for the separation of the United States from Europe by the Atlantic Ocean. The negotiations have illustrated the strain between the attempt to make parity militarily sensible, which involves expanding the concept to include all factors relevant to military performance in warfare, and the demands of negotiability which require narrowing it to the point where a formula can be found to demonstrate that parity already exists. Unfortunately even this has proved difficult, for NATO intelligence is unable to locate a limited area of real parity, while the Warsaw Pact, with different figures, claims it exists. NATO still insists on creating parity out of disparity; the Warsaw Pact offers it the option of fudging the numbers.

SALT and MFR both show the difficulties faced as arms control has moved into areas of serious asymmetries where a number of countries can claim to have a direct interest, where the relevant capabilities often defy ready verification and where there is no obvious, pre-existing East-West parity waiting to be ratified. The consequence has been, and is likely to be to a greater extent in the future, futile, intractable and bad-tempered negotiations.

Detente

This consequence is significant because it goes against what has generally been advertised as a major benefit of arms control in contributing to the 'atmospherics' of detente, as if the fact of discussion between potential enemies has been as important as the subject of the discussions. Arms control is represented as a gradually expanding area of co-operation, forcing contraction in the areas of conflict. Military detente, it has been argued, promotes political detente.

The most recent example of this has been the push in Western Europe in 1981, led by Chancellor Schmidt of West Germany, to get some talks with the East underway on the question of theatre nuclear forces. One aim which came to the fore in 1980, as East-West relations reached their lowest point for many years following the Soviet invasion of Afghanistan, was to provide a point of contact on one of the main issues that divided the two sides. The agreement reached at the Moscow Summit of July 1980 between Chancellor Schmidt and President Brezhnev was presented almost as a beacon of light in the general gloom over detente. It followed months of speculation over whether or not the Kremlin would move from its stand of not negotiating until NATO cancelled its decision to press ahead with the Pershing and Tomahawk missile programmes. Little of this speculation addressed the question of exactly what was to be discussed and to what end. The issue was presented as being simply one of an arms race versus negotiations.[12]

Once again the argument for successful negotiations has been used to support the act of negotiating as if the alternatives were only, as Churchill put it, to 'jaw jaw' or 'war war'. The trouble is that the content of the negotiations is ultimately more important in its political consequences than the form. If the only agreements to emerge are trivial then this may make only slight differences to the fundamental political relationship. If basic issues are tackled and the degree of antagonism between the two sides remains high, then agreements concerning central issues of national security will be difficult both to negotiate and then justify domestically. If the problems prove intractable then the result might well be a general deterioration in political relations. Frustrating arguments in SALT have raised third-order issues (Backfire-range, telemetry of missiles tests, cruise missile missions) into matters for discussion at the highest level. The consequence has been more acrimony than comity. The experience of SALT suggests that negotiations themselves are insufficient to prevent a worsening of relations. The dominant influence is the political climate not the negotiating climate.

The 'Fig-Leaf' Effect

The background to the discussions on theatre nuclear forces illustrates a second type of political objective, as powerful in practice if somewhat cynical. A compelling argument for entering into arms control negotiations is that in the absence of negotiations there may well be irresistible domestic pressure for unilateral reductions in capabilities. A major motive for entering into MFR was to head off 'Mansfieldism'. Once in negotiations any arguments for cuts or deferred deployments can be rejected as a unilateral concession. Withdrawing American troops from Europe could be presented as allowing the Russians to escape reciprocal obligations. The arms control proposal attached to the NATO decision to modernise long-range theatre nuclear forces was as much to placate those in Western Europe (particularly Holland and Belgium) nervous about this programme as out of any desire to make some deal with the Russians.

The negotiations themselves can be seen as a challenge to the unity of the West in that they provided the USSR with the opportunity to divide it by offering carrots to some and issuing threats to others. NATO initially managed to meet this challenge successfully and has learned to consult more in the process. Whether this will continue is, at this point unclear, given burgeoning anti-nuclear sentiment in Europe. Nevertheless, at least during the Carter Administration, Alliance cohesion was surprisingly robust.

The assumption was that the measures proposed by President Carter to reduce the risks of nuclear non-proliferation and arms transfers to the Third World would result in a clash with the allies whose commercial interests pointed them in the opposite direction. After some tense moments, the Carter Administration had to recognise the need to scale down its aspirations in these areas, particularly where excessive zeal might jeopardise alliance relations. Thus far, any enthusiasm for new patterns of international order, whether in the form of a deep detente with the Soviet Union or control of the diffusion of advanced military technologies, has not been allowed to interfere with the basic foreign policy objective of preserving a modicum of cohesion within the Western Alliance. At this juncture, however, the success of this policy of Alliance 'containment' remains fragile at best.

Assistance in fending off domestic opposition, in maintaining force levels, easing the way for new programmes or forging alliance solidarity, are rarely proclaimed as the objectives of arms control and true believers might see them as the negation of arms control. Yet these objectives, if rather negative, are quite real to the governments involved and have largely been achieved. The fact that so little has changed over the past decade may be more a source of satisfaction than regret to those actually engaged in many of the key negotiations.

Conclusion

What can be suggested from the experience of the 1970s?:

(a) Arms control cannot be a radical instrument. At best it can help conserve favourable and stable international relationships or military balances that have emerged independently. It does not offer a route to a new international order; rather an imperfect regulatory mechanism for the existing order. It cannot withstand adverse trends in political relations.

(b) The main effects of the various negotiations have been felt in foreign policy rather than defence policy, and the effects on the force structure have in general been indirect (highlighting particular gaps in capabilities or giving undue prominence to some systems) rather than direct (limitations, reductions, prohibitions). The few successes have been when arms control proposals have been in conformity with the contemporary military and technical wisdom.

(c) Arms control measures require a clear view of the sources of instability if they are to act as a corrective. When, as at the moment, there are many and often contradictory views on this matter, measures are extremely hard to design.

(d) Force structures are determined by a number of factors, of which the forces of the adversary are but one. It therefore cannot be assumed that an adjustment on one side will inevitably lead to a comparable reduction on the other, either as mutual disarmament or an arms race.

(e) Arms control theory must take account of the dynamics of negotiating between two or more nations aware of diverse interests and each with other policy objectives that are satisfied in preference. The more interests to be accommodated and the greater their diversity the harder it is to reach an agreement, and the more likely that if reached it will be fudged, untidy, ambiguous and irrelevant.

(f) The advantages of arms control as a diversionary move for hard-pressed defence planners or as an opportunity for ambitious diplomats are qualified by the problem of actually having to sit down and negotiate. There are many pressures to negotiate, for

> the best of motives and some more dubious, but these are rarely matched by either the will or the wherewithal to bring these negotiations to a successful or domestically popular conclusion.

There is little wrong with the principle that, in this dangerous age, Superpowers (and their allies) should talk regularly and in detail about their respective force structures, nor with the hope that they can take remedial action to remove sources of disturbance in their strategic relationship. The disillusion and frustration surrounding arms control results from a pressure to negotiate for its own sake and then to produce regular agreements. Political harm has been done by being seen to be in disagreement and little of strategic relevance has been achieved. There was, in reality, not a lot to do (particularly in the nuclear area) and what could be achieved could neither be justified nor explained by reference to any strategic doctrine but only to its 'negotiability'. The essence of arms control theory, that potential enemies can co-operate in the military sphere, has been discredited.

There are grounds for believing that some enthusiasm for arms control will return. Sections of public opinion, particularly in Europe, are resistant to the notion that there is no diplomatic means of improving East-West relations. The extent to which President Reagan was forced to retreat from his hostility to SALT during his election campaign is instructive. The cost of armaments will continue to grow and, in recessionary times, there will be pressures for cuts. Inflation is the Great Arms Controller. Apart from anything else the current negotiations have yet to be abandoned and neither side appears anxious to take the responsibility for concluding the exercise.

Reformers always want to start from scratch, to rebuild from the foundations. This is rarely possible and the starting point for a new policy must be the current situation. In SALT, we have now travelled so far down the long and winding road to parity that some means must be found for agreeing that a satisfactory destination has now been reached. The Soviet leadership was content with SALT II but must now consider conceding the principle of amendment to a signed treaty so that President Reagan can claim that it is no longer 'fatally flawed'. Some adjustments may be available here, but it is disturbing to find the new Administration hinting at an even better SALT Treaty to come, indicating a restlessness that may be harbinger of more years of contentious diplomacy for slight effect.

It is also disturbing to find the theatre nuclear force negotiations already moving down this long and winding road. The best idea might be to merge these new talks with SALT III[13] so as to develop a collection of ceilings and sub-ceilings that might stay in force for some time. This would recoup some of the investment put into arms control and provide a tolerably predictable strategic environment.

The important thing is to know when to stop. Like compulsive gamblers the experience of failure and frustration has been disregarded in a drive for a big success. Rather than continue to aim for some new arms control regime of immense complexity and profound strategic consequences, it might be better to follow a less demanding path.

My argument is to reorganise SALT and MFR into unashamed 'talking shops', one for nuclear and the other for (European) conventional arms. The discussions should take place at the highest political and military levels on a regular basis. There is something of a precedent for this sort of discussion in the conferences of 'experts' established to consider the test ban and, less successfully, surprise attacks in the late 1950s, as well as in the preparatory commissions for the set-piece disarmament conferences of the inter-war years. The objective for these discussions should be to clarify the nature of the interaction between the forces of each side, to note each side's anxieties and phobias and possible means for calming them, and to identify new types of capabilities that could have a destabilising impact. If problems susceptible to negotiated agreements should arise then small sub-committees could be set up for this purpose. It is essential that a series of formal agreements is not seen as the overriding purpose. It may be possible to suggest areas where unilateral (but not irrevocable) actions may be beneficial.

I would envisage that the areas where modest agreements might be possible would be similar in kind to the ABM Treaty, constraining potentially disruptive, but not as yet fully exploited, developments. Examples might be controls on anti-satellite weapons or depressed-trajectory missiles. There might be possibilities for confirming or adjusting the various SALT ceilings. In discussions on conventional forces in Europe it is to be expected that 'confidence-building measures' (CBMs) will come to the fore. These have the advantage of dealing directly with sources of concern over surprise attack, but have yet to move to a level where moves to launch a surprise attack could be seriously impeded.

There are a number of objections to this limited approach. The proposal is institutionally ambitious, in that it deviates from traditional forms of diplomatic activity in this area, yet promises little concrete by way of results, and therefore fails to get to grips with the dangers inherent in the East-West military competition. However, I consider the military relationship to be tolerably stable and therefore in itself not requiring major adjustment. There are areas where the politico-military establishments could usefully be reassured, or make clear their concerns, over the meaning of specific trends in the other side's force structure.

A more difficult objection is that the proposal may reflect a liberal faith in the positive benefits of communication and information. This is a risk but it is one that faces any arms control effort, as the detractors are quick to point out. It requires a willingness to be frank and open and a readiness to discuss detailed aspects of military deployments. This does not come easily, particularly to the Russians, but a sensible recognition of how much is now known by 'national means' has already encouraged greater frankness. However, as we have seen in MFR, it is by no means clear that greater openness leads to greater trust and understanding. Without the pressure to negotiate, it may be possible to avoid essentially empirical judgements becoming too politicised. Nevertheless the problem remains that as more is found out about 'the potential enemy', the sense of 'potential enmity' can grow rather than recede.

A more basic charge is to assume that the Soviet Union shares the West's interest in stability. I do believe that there is a coincidence of interests here, although politically the two sides have many opposing interests, and even when agreeing on the need for stability, they have different views on the sources of instability and indeed on the character of the overall strategic relationship. One objective of simply talking is to clarify and illuminate the differences rather than attempting to create an artificial unanimity.

In the long term, radical measures may well be necessary for we cannot go on forever deriving our security from a balance of terror. Ideas such as those for banishing short-range nuclear weapons from the middle of Europe have their attractions, but it would be unwise to push them too hard at the moment. Rather we need to use the current conditions of strategic stability and a sour political atmosphere to prepare ourselves for the opportunities that might arise when political conditions are substantially improved.

Unfortunately, 'military detente' cannot precede or promote 'political detente'. The military relationship reflects, and can reinforce, trends in the political relationship but it is not overriding. The reason for modesty in arms control for the moment is the unfavourable political climate. When it improves more may well be possible, but if we are to avoid the mistakes of the past, arms control must not then be loaded with too many political burdens.

Notes

1. Some of the considerations raised in this chapter are also dealt with in my Chatham House Paper, *Arms Control in Europe* (RIIA, London, 1981).
2. Two of the seminal texts were Thomas Schelling and Morton Halperin, *Strategy and Arms Control* (The Twentieth Century Fund, New York, 1961) and Hedley Bull, *The Control of the Arms Race* (Praeger, New York, 1961).
3. See for example, Alva Myrdal, *The Game of Disarmament: How the United States and Russia Run the Arms Race* (Manchester University Press, Manchester, 1977).
4. The 'comprehensive' SALT proposals taken by Secretary of State Cyrus Vance to the Soviet Union in March 1977 represented a consensus in Washington but were unacceptable in Moscow. The comprehensive proposals then became a standard by which opponents of concessions to the Soviet Union judged the Carter Administration as it sought some compromise SALT agreement.
5. Robert Osgood, *Limited War: The Challenge to American Strategy* (University of Chicago Press, Chicago, 1957).
6. Phil Williams, *Crisis Management* (Martin Robertson, London, 1976).
7. Kissinger once confided: 'I would say in retrospect that I wish I had thought through the implications of a MIRVed world more thoughtfully in 1969 and 1970 than I did. What conclusions I would have come to I don't know.' Background briefing by Henry Kissinger, 3 December 1974, reprinted in *Survival*, vol. XVII (July-August 1975), p. 194.
8. This had been assumed to be imminent, and thus began to influence policy, as early as 1969. It took until the second half of the 1970s when the Soviet build-up of MIRVed ICBMs actually took place, for it to become a matter of general concern. See Lawrence Freedman, *US Intelligence and the Soviet Strategic Threat* (Macmillan, London, 1977).
9. There is serious concern in the US Air Force as elsewhere that the M-X basing concept of multiple aim points could fail to provide the necessary vulnerability without the sort of limits imposed on the proliferation of accurate warheads by SALT II.
10. See Colin Gray, *The Soviet-American Arms Race* (Saxon House, Farnborough, Hants, 1976).
11. The full title is 'Mutual Reduction of Forces and Armaments and Associated Measures in Central Europe', which is a mouthful even as an acronym.
12. The problems here are discussed in Lawrence Freedman, 'The Dilemma of Theatre Nuclear Arms Control', *Survival* (January-February 1981).
13. See Ibid.

3 ARMS CONTROL AND WESTERN SECURITY: A QUESTION OF GROWING IRRELEVANCE[1]

Richard Burt

Although it is an exaggeration to argue that arms control has failed, it is clearly in trouble. A new phase of Superpower tension, the product mainly of Soviet activism in the Third World and American inconstancy, also seems to have marked the end of an era of East-West negotiations over controlling military forces. Negotiations in some areas, such as limiting NATO and Warsaw Pact manpower levels in central Europe, are continuing, but arms control, for the present at least, no longer forms part of a wider process of political accommodation. Although the Carter Administration struggled mightily to insulate the Strategic Arms Limitation Talks (SALT) from other facets of the Soviet-American relations, this heroic (if politically naïve) effort has failed. While still resting on the Senate calendar, action on ratifying the long-delayed agreement was deferred indefinitely after the Soviet invasion of Afghanistan. Without SALT, other arms control enterprises, namely efforts to conclude a comprehensive nuclear test ban and to control the testing and deployment of anti-satellite weapons, are also in eclipse. Meanwhile, concern in Washington over Western military weakness in the region surrounding the Persian Gulf ended President Carter's interesting and controversial attempt to engage Moscow into a dialogue over conventional arms transfers and the deployment of naval forces in the Indian Ocean.

It is thus tempting to conclude that a series of disruptions and possible misunderstandings in Soviet-American relations, capped off by Afghanistan, are mainly to blame for the current malaise in arms control. And it is equally tempting, then, to argue that if the two sides are able – in one, two or five years – to re-establish a working political dialogue, that arms control will experience a renaissance. It is undoubtedly true that detente, during the 1970s, was a political prerequisite to progress in arms control. But this does not mean that if the political conditions for Soviet-American negotiation become ripe in the coming decade that arms control is likely to once again become a chief avenue for exploring solutions to international security problems. Indeed, as this chapter will attempt to show, it is highly unlikely that arms control, whatever its salience for the Soviet Union, will play a major role in assisting the United States and its allies in addressing a new range of military problems in the 1980s. To accomplish this aim, four questions will be addressed:

1. What security problems is the West likely to encounter in the coming decade, and how relevant will arms control be to their solution?
2. What basic assumptions about military competition underlie the negotiating process and how applicable are they to the strategic environment of the 1980s?
3. What is the relationship between current arms control practice to past arms control theory?
4. What unilateral options for solving military problems will be available to the West in the 1980s and how should these options be integrated with a strategy for arms control?

Arms Control and the Emerging Security Environment

Although frequently given to hyperbole, Henry Kissinger was probably right when he asserted recently that the United States in the 1980s was moving into as perilous an era as any in its history. For the foreseeable future, the United States and its allies will be forced to contend with security challenges that are unprecedented both in scale and complexity:

(i) Although the threat of using nuclear weapons has been the bedrock of Western deterrence for 30 years, it will be difficult in the 1980s for the United States to maintain the existing conditions for parity in strategic forces with the Soviet Union. Spending somewhere in the vicinity of two-and-a-half times more on strategic forces than the United States, the Soviet Union possesses a momentum in offensive and defensive systems that should give it an advantage in most, if not all, indices of strategic capability (missile throw-weight, equivalent megatonnage, total warheads) by 1985. More important, the most reliable and responsive component of the American strategic 'triad' – the Minuteman ICBM force – will become vulnerable, in theory, to a Soviet disarming strike. Although the United States will retain an undisputed 'assured destruction' capacity against the Soviet Union through the decade, the implications of an eroding balance in both static and operational capabilities for Soviet behaviour and Western political and military cohesion could be enormous.

(ii) In the European theatre, the traditional Soviet advantage in conventional (and chemically-armed) forces is now being complemented with a new family of nuclear weapons, such as the Soviet SS-20 IRBM and the Backfire bomber. Responding to these developments, the Alliance is tenuously moving to deploy a new generation of land-based ballistic and cruise missiles in central Europe. But the SS-20 and Backfire are only the most visible of a comprehensive programme of nuclear modernisation which includes new nuclear-capable aircraft and several new nuclear battlefield support missiles. As in the area of long-range forces, Western options are gradually being narrowed down to relying on the threat of massive escalation, an alternative in peacetime that is likely to lack credibility and in wartime is likely to be viewed as suicidal.

(iii) In other regions, the growth of Soviet power projection capabilities will coincide with growing Western dependence on resources in the Third World, particularly oil in the Persian Gulf. In some regions, perhaps Southeast Asia, the West will be able to rely on continuing Sino-Soviet hostility to deter any Soviet threats to Western access. In other areas, such as the Middle East and the Persian Gulf, the West, mainly the United States, will be obliged to act as an outside guarantor of stability. Yet the problem of defending against Soviet military intervention or large-scale local conflicts that could jeopardise Western oil supplies are enormous: the political prerequisites for establishing a system for collective security in the region are absent and the Soviet Union enjoys major geographical advantages in moving forces into the area. The deployment of American naval forces in the Indian Ocean, and later on, the existence of a 'rapid deployment force', will give the West something like a 'trip-wire' defence, but that, at best, could deter widescale fighting in the region but, at worst, might only trigger a large-scale military disaster.

(iv) The challenges posed by the expansion of Soviet military power into the Third World will be exacerbated by regional violence and decay. In the 1980s, the West will not only have to contend with persistent regional rivalries and disputes, such as the Arab-Israeli dispute; there are also signs that economic strains, coupled with cultural and political disruptions, will lead to the disintegration of existing rule in countries such as Turkey. The strategic problem in this context is not so much that the Soviet Union, bent on conquest, will decide to intervene with military force – the 'World War II' model. It is that both Superpowers, possessing conflicting commitments and differing perceptions of how rapid change in the Third World could affect their interests, could find themselves swept into a series of confrontations in unlikely locales – the 'World War I' model.

How successful will arms control negotiations be in addressing these challenges? Of course, it is probably asking too much from arms control to expect it to cope with a First World War scenario in which Moscow and Washington find themselves at the brink of war after becoming embroiled in a regional conflict. Formal 'confidence-building measures', like the 'hot line' established in the 1960s between Washington and Moscow, can reduce uncertainties in time of crisis, but the management of severe tension requires human skills rather than technical fixes. All the same, there are strong reasons for believing that even in the more traditional realm of East-West military competition, arms control negotiations are likely to be of little utility in the coming decade.

To begin with, despite the best hopes of arms control proponents in the West, existing negotiations have had little discernible impact on Soviet thinking about military forces. Accordingly, whatever else negotiations at SALT and in other forums have produced, arms control, during the 1970s, failed to produce a shared consensus between the two sides over the role of military power, in particular, how various weapons deployments affected the stability of the overall military balance. Without such a consensus, negotiators have failed to work out solutions to such problems as the emerging vulnerability of land-based ICBMs. For instance, since Albert Wohlstetter's path-breaking Foreign Affairs contribution in 1957, it has been an article of faith within the Western strategic community that both the United States and the Soviet Union should avoid taking military steps that could threaten the survival of the other side's strategic forces. In this way, it was argued, neither side would have strong incentives for pre-empting in time of crisis. Since 1969, at SALT, the United States has championed controls on weapons, such as anti-ballistic missile defences, that were understood to threaten 'strategic stability'. In its famous March 1977 proposal, for example, the Carter Administration tabled a proposal that, from the traditional arms control perspective, would have restructured both sides' arsenals along 'safer' lines: the most threatening (and vulnerable) component of both sides' forces – fixed-site ICBMs – would have been de-emphasised while each would have been free to build up its more secure sea-based and bomber forces.

While a host of explanations have been offered to explain Moscow's rejection of the proposal, the Soviet strategic programme during the last 15 years offers strong support for the conclusion that if Moscow does possess a concept of 'strategic stability', it is not the one that has guided American defence policy or negotiating strategy at SALT. It is futile to debate whether, as some writers have suggested, Moscow is acquiring the capacity to 'win' a nuclear war. Instead, it is sufficient to note that rather than mirroring Western concepts of deterrence, Soviet strategic programmes, from the deployment of counter-silo ICBM forces to the orbiting of anti-satellites, tend to reflect the preference for nuclear 'war-fighting' found in Soviet literature. What the experience so far at SALT reveals, along with the record at the MBFR talks in Vienna, is that lacking any agreed notion among negotiators over what stability constitutes, the most likely outcomes of arms control are agreements like the SALT II treaty – accords that ratify rather than restructure the prevailing character of the military balance. This is clearly one important reason why the Vienna talks continue to be stalemated; while there are sound political reasons for the West to insist on manpower parity in the centre region, the Soviet Union possesses equally strong incentives for maintaining its existing superiority.

Of course, doctrinal convergence is not a prerequisite for arms control agreements and nations sometimes enter accords that run contrary to their doctrinal proclivities. Even though the Soviet Union seems to place great emphasis on limiting damage in nuclear conflict, Soviet negotiators, in 1972, finally agreed to severe restrictions on antiballistic missile deployment. But this was hardly because Soviet planners, as some Western observers argued at the time, were persuaded that 'mutual nuclear vulnerability' offered the only rational basis for strategic force design. Instead, a much more plausible explanation for Moscow's action was that with the American Safeguard ABM system about to go into deployment, a hard-headed decision was taken to close off competition in a realm where the United States was seen as possessing important technical advantages.

The major question now is whether the United States will possess the leverage in the 1980s to force Moscow to take similar actions. It is hazardous to make predictions, but it is highly doubtful. Still in the process of deploying its fourth generation of ICBMs (the SS-17, SS-18 and SS-19), the Soviet Union is likely soon to begin testing a series of four or five new land-based missiles, some of which could conceivably be deployed in a mobile mode. At the same time, Moscow is also said to be developing two new strategic bombers. At sea, the Soviet Union is deploying a new long-range SLBM, the SS-N-18, aboard its Delta-class SSBN and is testing another SLBM, the Typhoon, for deployment aboard a new class of submarines. In the defensive area, the Soviet Union's gradual augmentation of its civil defence complex continues; development of anti-submarine warfare forces is accelerating; the modernisation of Moscow's ABM system is reportedly underway and testing of anti-satellite systems, after an eighteen month lull, has started again.

After a period of limited growth, the United States is now preparing for a new period of strategic modernisation, at least in the area of offensive forces. The navy's new class of Trident SSBN, with the long-range C-4 missile, have started deployment and the air force, in 1982, will begin receiving long-range cruise missiles for deployment aboard the aging B-52 force. The air force will also soon begin to modernise the Minuteman force with a higher-yield warhead, the Mk12A, and started engineering development of a land-mobile ICBM, the MX. It is unlikely, however, that these programmes will provide American negotiators with the necessary bargaining capital to achieve arms control solutions to such pressing problems as ICBM vulnerability. Through most of the 1980s, the United States will lack the means of matching Moscow's capacity to threaten American ICBMs. With the MX, which is scheduled to be fully deployed in 1989, the United States will possess a system capable of threatening a large fraction of Soviet ICBMs and, in theory at least, of surviving any Soviet threat. But it is hard to see how the MX could be used as a 'bargaining chip' in any future SALT negotiation. While Moscow might be willing to agree to accept a permanent ban on the deployment of mobile missiles in order to stop the MX, in such an accord the Soviet Union would continue to retain its capacity to threaten fixed-site American missiles. Thus, in order to address this problem, American negotiators would need to insist that in order to give up the mobile missile option, the Soviet Union would have to dismantle a good portion of its existing fixed-site missile force. As desirable as such a proposal would be from Washington's perspective, it would hardly be negotiable from Moscow's.

Similar problems would be likely to exist in any future negotiations covering theatre nuclear forces in and around Europe. A basic rationale for the December 1979 decision by the Alliance to deploy 108 Pershing II ballistic missiles and 464 ground-launched cruise missiles in central Europe was to provide Western negotiators with the necessary bargaining leverage to limit further Soviet deployment of the SS-20 and other forces targeted against Western Europe. But even with the new systems, the West is almost certain to be at a major disadvantage in the theatre nuclear realm in the mid-1980s. With the Soviet Union in the process of deploying several new nuclear-capable aircraft and missiles targeted against NATO, it is almost impossible to envision a negotiating proposal that would be attractive to the West while standing a chance of being accepted by Moscow.

Even if Western negotiators find themselves with enough leverage to gain Moscow's interest in new agreements, arms control negotiations are still likely to get bogged down for other reasons. As Leslie Gelb has pointed out, as negotiators at SALT search for a more comprehensive agreement to replace the constraints placed on land and sea-based launchers in 1972, the negotiations have grown infinitely more complicated.[2] In 1979, a technical disagreement over the extent to which the Soviet Union was to be allowed to encode missile telemetry almost derailed the entire SALT II treaty. In the future, disagreements over technical issues associated with verification or other issues, are likely to grow more common. For example, a favoured approach for addressing possible restrictions on theatre nuclear forces is to discuss them within the framework of SALT. But if, as suggested above, a desirable outcome to limiting theatre nuclear forces is non-existent, then the next phase of SALT would become a hostage to disagreements on this issue alone. As Leslie Gelb has suggested, one solution is to simplify arms negotiations. The problem with this proposal is that in the complicated security environment of the 1980s, simply-drawn arms agreements are unlikely to be of much military value. In fact, this seems to be the fate in store for the MBFR exercise. While perhaps important politically, an agreement on manpower levels in central Europe would probably have no discernible effect on the capacity of the Soviet Union to launch a 'short-warning' blitzkrieg against the Alliance.

As the MBFR case illustrates, perhaps one of the salient characteristics of arms control in the 1980s will be its irrelevance. Having failed to bring about stylised conceptions of 'stability' elaborated by Western thinkers two decades ago, the East-West military balance is evolving to a point where operational effectiveness, as opposed to overall force size, is emerging as the key factor in assessing the balance.

Thus, the Soviet Union's key advantage in Western Europe lies with its 'combined arms' approach to land warfare which enables Soviet forces to orchestrate firepower and manoeuvre elements to cope with NATO improvements. But the Soviet 'combined arms' approach can only indirectly be influenced through negotiations. In the area of strategic forces, a similar phenomenon is at work. In the 1960s and 1970s, quantitative differences in American and Soviet forces – both in overall size of arsenals as well as individual weapons' attributes – were seen as key to judging the capabilities of the two sides. But in the post-parity era, other factors will take on greater importance. In the 1980s, the most important elements of the balance in long-range forces will be those that relate to force effectiveness – target acquisition; command, control and communications (C^3); force redundance, survivability and endurance. These attributes, unlike missile launchers, warhead numbers and throw-weights, are difficult to measure and are thus less amenable to control in negotiations. As C^3 and other factors of force effectiveness take on greater importance in judging the capabilities of the two sides, it may become necessary to distinguish between two, separate strategic balances: a *symbolic* balance, based on static hardware counts, that is of primary interest to the arms control community and an *operational* balance that represents the real capacities of the two sides to engage in nuclear conflict.

Arms Control As A Problem

While negotiations in the 1980s appear unlikely to grapple with the major military challenges confronting the United States, arms control could create a variety of obstacles to surmounting them unilaterally. The problem is not simply that arms control is irrelevant, but that current negotiating approaches are based on assumptions about East-West military competition that, if once accurate, are no longer valid. It is now a cliché that arms control cannot serve as a substitute for an adequate defence posture, but in many respects, Western negotiating strategy in many areas seems to assume that it can. Indeed, some of the basic beliefs that underpin Western thinking about arms control seem to rule out the possibility of achieving greater compatibility between negotiating policy and military strategy.

Three fallacies of arms control, in particular, are worth mentioning. The first is that arms competition can, and should, be frozen by agreement. The notion that there is some perfect equilibrium in the military state of affairs that can be defined and sustained through agreement is a recurring one in history. In the early 1930s, for example, some of the best minds in Europe attempted to define ways of bolstering military stability by prohibiting the deployment of 'offensive' weapons. But as American and Soviet negotiators discovered in the 1970s, 'stability' is an elusive concept. While British experts maintained that the tank was an 'offensive' weapon and should be controlled, French military planners saw them as primarily defensive systems. But notions of 'stability' do not merely depend on one's geostrategic situation. Definitions of 'stability' are also highly dependent on the existing state of technology and the relative state of the balance. For example, in the early 1960s, when the United States possessed a credible counterforce capability against the small Soviet long-range force, Secretary of Defense Robert McNamara's concept of a 'no cities' nuclear strategy seemed a stabilising option in the sense that, together with the larger American civil defence and bomber defence effort, it provided few incentives for the Soviet Union to undertake a nuclear strike. Toward the end of the decade, however, McNamara himself recognised that a damage-limiting strategy had been ruled out by the growth of Soviet strategic forces and proposed instead that 'stability' should reside mainly in the capacity of the two sides to threaten each other's largest cities. In advocating a 'mutual assured destruction' strategy, McNamara strongly resisted the deployment of ballistic missile defences, a position that was codified in the 1972 ABM treaty.

Although strongly condemned by Donald Brennan, among others, it is at least arguable that the ABM treaty contributed to stability within the military environment of the early 1970s, when neither side's offensive retaliatory capacity was under threat. But in an era when American ICBMs are becoming vulnerable to Soviet attack and when Moscow is engaged in a sustained effort to bolster its strategic defences, the contribution of the ABM treaty to 'stability' has become questionable. There is still a case to be made for prohibiting either side from defending urban areas with ABMs, but the constraints within the ABM treaty on protecting land-based ICBMs and other military facilities have foreclosed an important solution to the Minuteman vulnerability problem in the coming decade. Either deployed around the existing Minuteman force or, more effectively, in conjunction with a multiple-aim-point MX system, point-defence ABMs are likely to offer the most cost-effective means of countering the Soviet counter-silo threat. An arms control accord that might have been conducive to stability in one phase of the strategic competition has become positively counterproductive in another.

A second fallacy is that new weapons that threaten arms control negotiations somehow also pose a danger to military stability. This fallacy can be seen to have operated in the case of the cruise missile at SALT II. It is not hard to understand why the cruise missile so severely complicated efforts to complete a new SALT treaty. The relatively small size of the missile and its compatibility with different launcher vehicles posed significant challenges to prevailing verification procedures. More important, the missile's potential utility both as 'strategic' and a 'tactical' weapon created daunting problems of categorisation. Thus, there were strong incentives for American negotiators to find a formula for limiting land and sea-based versions of the system in the Protocol to the Treaty while agreeing to deploy air-launched cruise missiles on heavy bombers alone. The irony, of course, is that while cruise missiles posed a threat to the SALT process, their deployment, particularly in theatre roles, would more than likely strengthen deterrence. Clearly a second-strike weapon in a strategic role against the Soviet Union, their ability to offer more survivable basing modes at land and sea and their possible marriage to a new family of conventional munitions would clearly bolster regional defences, in Western Europe and in other areas.

New weapons technologies, of course, are not always desirable. In some cases, new systems can increase first-strike incentives and unsettle the political *status quo*. In the longer term, they can also force wrenching changes in prevailing patterns of military thought, as space-based laser ABM systems might necessitate in the 1990s. At the same time, new weapons technologies can sometimes help governments attain the three classic objectives of arms control: enhancing deterrence, shoring up defence if deterrence fails and saving money. Had the SALT process been underway in the early 1960s, the Soviet Union, together with many arms control proponents in the United States, would have probably maintained that the deployment of Polaris SSBNs created verification problems and represented another 'destabilising' round of the so-called 'qualitative arms race'. It did represent a new departure in military technology, but in a period of ICBM vulnerability the deployment can hardly be viewed as anything but a beneficial development. If arms control is not a substitute for unilateral military initiatives, then the costs for negotiations of deploying new systems must be measured against the military benefits that will be derived from their deployment.

The cruise missile case is also useful in highlighting the third fallacy of arms control: the belief that negotiating boundaries reflect military reality. As the acronym implies, the SALT process uses as its central organising principal the idea that there is a distinct class of Soviet and American 'strategic arms'. While this may have been a generally accurate description of the state of technology in the 1960s, a new class of more flexible weapons, like the cruise missile, is making traditional distinctions between 'strategic' and 'tactical' arms obsolete.

As a result, the SALT process, with its preoccupation with a mythical 'homeland-to-homeland' nuclear balance between the Superpowers, has become an inappropriate and, for the West, a politically damaging forum for coping with weapons systems that are relevant to both intercontinental and regional warfare. SALT outcomes that limit these weapons are unattractive because they foreclose options for upgrading theatre defences. At the same time, agreements that exclude these systems are equally unattractive, because, as the controversy over the Soviet Backfire bomber illustrates, both sides would be able to increase their intercontinental-range arsenals under the guise of expanding their theatre forces. Because of the American nuclear commitment to Western Europe, the concept of a 'homeland-to-homeland' balance fostered by SALT has always been pernicious. But during a period in which the systems controlled by the process were not directly relevant to the defence of the European theatre, bilateral agreements covering long-range forces were bearable. In a period, however, when it is becoming increasingly difficult to disaggregate the Soviet-American strategic relationship from the wider East-West military balance, new SALT agreements could endanger the military interests and political cohesion of the Western Alliance. Despite the superficial attractions of moving toward a more comprehensive negotiating forum, this problem will not be solved by moving, as the Alliance seems bent on doing, to incorporate Soviet and American long-range theatre nuclear systems in negotiations in the next phase of SALT. While discussing long-range theatre systems at SALT would require negotiators, for the first time, to rigorously consider the implications of various negotiating outcomes for regional defences, it could also foster an analytical decoupling between longer-range theatre weapons and shorter-range systems deployed in and around Europe.

If SALT points out the dangers of trying, for purposes of negotiation, to isolate a functional category, other arms control exercises underline the problems of geographical compartmentalisation. One of the principal drawbacks of the MBFR negotiations, for example, is the geographical asymmetries inherent in efforts to limit forces in an artificially-bounded chunk of central Europe. As an adjacent land power, the Soviet Union possesses vast advantages in projecting additional troops and firepower into the area. In a similar war, military reality was distorted by the geographical boundaries erected in the now-defunct Indian Ocean negotiations launched by the Carter Administration in 1977. Freezing American and Soviet forces at sea while not imposing similar constraints on land-based maritime capabilities would have clearly worked against Western interests in South West Asia. While limiting the deployment of American carriers in the region, the Soviet Union would have been free to expand numbers of Backfire bombers and other land-based naval strike aircraft that could be used in the area.

Rediscovering Arms Control and Revitalising Defence Policy

In retrospect, it is curious that in the West at least, arms control and defence planning seem to be working at cross purposes. Arms control, as a body of thought, developed in the late 1950s and 1960s as a realistic response to the obvious failure of classical disarmament initiatives, such as the Baruch Plan in 1947, to bring an end to East-West military competition. As a starting point, the original arms controllers of two decades ago accepted the fact that the 'arms race' could not be brought to an end without an amelioration in East-West political competition. Thus, the goal of arms control, expressed perhaps most eloquently by Hedley Bull 20 years ago, was not necessarily to reduce military forces but to restrain their use in a world in which East-West competition would continue. In this way, Bull argued that 'the promotion of international security is of course a matter of unilateral armaments policy as well as arms control. Military situations favourable to general international security may be brought by unilateral action, whether this takes the form of rearmament, disarmament or other forms.'[3] Bull's point, of course, was that arms control, defined as 'restraint internationally exercised upon armaments policy' was only one of a variety of means available to governments for seeking a more secure military environment.

Over the last 20 years, the notion of arms control as an instrument, as distinct from a competitor, to military policy has developed from a theoretical pastime into an institutionalised activity. As a result, arms control has developed the same kind of mindless momentum associated with other large-scale, bureaucratic pursuits, and conceptual notions of limited durability have gained bureaucratic constituencies and have thus been prolonged beyond their usefulness. In the end, the goal of simply obtaining an agreement has tended to supplant earlier security objectives. A basic rationale for pursuing force reductions in Europe, for example, was to dampen enthusiasm in the early 1970s in the Senate for unilateral withdrawals of American troops. That rationale has ceased to exist, but the MBFR dynamic continues. Institutional and conceptual inertia in arms control was also apparent in efforts by the Carter Administration, in 1977, to ban the deployment of mobile ICBMs, despite the fact that the Air Force's MX had emerged as the principal option available to the United States in coping with the imminent vulnerability of the ICBM force.

More generally, the arms control debate has tended to revolve around obsolete conceptions of the Western security problem. Although the Soviet Union, over the last decade, has demonstrated its lack of devotion, or at least inattention, to arms control goals developed by Western analysts in the earlier 1960s, much of the current debate seems to assume that decisions by the United States will have a major impact on Soviet actions. Thus, the current debate over whether the United States, with the MX or a similar system, should acquire the capacity to threaten the bulk of Soviet land-based missiles has proceeded on the assumption that Washington can dominate the course of arms competition. But Moscow is already acquiring the capacity to threaten American missile silos. The choice implicit with the MX is not whether the Superpowers should retain a system of 'mutual deterrence' with the Soviet Union based on only targetting urban and industrial concentrations; the choice for the United States in the 1980s is whether it wants the Soviet Union to be the only Superpower to be capable of destroying a large fraction of its adversary's strategic assets.

But it is too easy to blame arms control alone for the problems that the West is having in adjusting to a new strategic environment. If arms control should serve as a flexible instrument of military policy, then military objectives need to be spelled out with more clarity than is now the case. The problem of outlining a military strategy for the United States and its allies in the 1980s is clearly enormous. Unlike earlier periods, there is today no general national security doctrine, such as 'containment', to organise defence policy around. Nor do military planners possess the luxury of strategic superiority with which to avoid difficult choices. Instead, in an era of unprecedented Soviet military growth, the United States, still recovering from the shock of Vietnam, is profoundly uncertain over the meaning of the Soviet build-up and how to respond to it. Arms control, by promising to cope with the Soviet challenge while also restraining unnecessary military responses by the United States, has seemed to offer an opportunity for escaping a divisive and enervating debate over defence policy.

But if the thesis of this chapter is correct, in order for arms control to play a useful role, much of the confusion over American goals in such areas as strategic doctrine, the relationship between theatre nuclear and conventional forces, as well as maritime and projection capabilities, must be outlined with far greater precision than is now the case.

It is possible to suggest some of the attributes of an American defence policy for the 1980s. As appealing as it sounds to argue that the United States can respond to the growth of Soviet military power by emphasising either long-range nuclear forces or local defences, the era in which the West possessed such choices is gone forever. Out-spending the United States in both nuclear and conventional forces, the Soviet Union, over the last 16 years, has acquired a well-balanced and multi-faceted capacity to wage war. Although emphasising conventional forces will tend to raise the 'threshold' in local conflicts for the Western use of nuclear weapons, a conventional-emphasis strategy could actually provide the Soviet Union with incentives to escalate in time of war.

The Soviet buildup thus means that local military balances and the Soviet-American strategic relationship have become more interdependent. Rather than subdividing the overall East-West military relationship into artificial compartments, American planners will have to investigate new ways in which long-range and theatre forces can be integrated to respond to military threats across a wide spectrum. A mutually-reinforcing strategy for theatre and long-range forces would enable the United States to get away from relying on the self-defeating dichotomy between deterrence and defence. The threat of escalation will continue to provide both sides incentives for exercising restraint in local conflicts. But the degree of Soviet restraint will depend, in large part, on American possession of non-suicidal options for escalation. Accordingly, a new emphasis must be placed in generating nuclear responses that are militarily meaningful. 'Parity', a political definition of force sufficiency, is not an adequate measure of military effectiveness, because it fails to provide any operational requirements for long-range forces. 'Flexibility' is likewise a poor measure of effectiveness, because limited nuclear options, by themselves, have a suicidal quality. In the 1980s, 'strategic sufficiency' will depend on more than numbers and flexibility; it will require increasing attention to be paid to force responsiveness, survivability, and sustainability. It is to be hoped that the doctrinal adjustments announced in August 1980, and embodied in PD-59, will be matched by force planning to fulfill those objectives.

These qualities provide some indication of how some of the tough issues concerning American strategic forces should be answered in the years ahead. For a start, the requirement of military effectiveness necessitates the acquisition of secure and survivable forces capable of threatening a range of military forces, including hard-targets in the Soviet Union. At present, the MX appears to be the best candidate for this role, but there are numerous reasons to believe that the system may never be built. If the primary concern over the MX is its potential vulnerability to a surge in Soviet warheads, then technical options, such as marrying the system to ABM defences, should be seriously examined.

(These concerns are in evidence in the strategic attitudes of members of the Reagan Administration.) If these are found wanting, the United States should hedge its bets on the MX by making sure that the Trident D-5 SLBM will be available far sooner than its existing IOC of 1990. Whatever happens, discussions over future American force posture should not become bogged down in a search for perfect technical fixes. In addressing strategic survivability, there is a real danger that the best could become the enemy of the good.

On the defensive side of the ledger, there are a number of activities that need to be taken more seriously, including civil defence, antisubmarine warfare and air defence. In space, the potential offered by new technologies, particularly lasers, needs to be examined concurrently with programmes to reduce the growing vulnerabilities of American early warning, surveillance and communications spacecraft.

What Role for Arms Control?

As we have seen, unilateral force improvements, not arms control, are likely to offer the best solutions to the problems that will preoccupy Western planners in the years ahead. But arms control agreements can create a degree of predictability that is useful both politically and militarily. In criticising arms control, it is easy to lose sight of the uncertainty that plagued political leaders and military planners during the first half of the 1960s. In the early part of that decade, the tendency to exaggerate Soviet strategic programmes led the United States to rush ahead with programmes that by 1965, seemed unnecessary. However, American reactions in the early 1960s probably fostered a tendency in the latter half of the decade to underestimate Soviet strategic ambitions, an even more dangerous development. Whatever else SALT does, it should enable political and military authorities to agree on the character of the evolving strategic environment. The environment is clearly becoming more dangerous, and consensus over how the balance is shifting should enable policy-makers to agree on a programme for correcting deficiencies in the strategic posture.

But 'optical parity' in launcher numbers (to use Colin Gray's term) and predictability are not the same things as the maintenance of deterrence. In strategic forces, the maintenance of deterrence requires the deployment of American land-based systems in a more survivable basing mode and, more controversially, an enhanced ability to threaten hard-targets in the Soviet Union, in the opening and later phases of a conflict.

In the European theatre, deterrence also requires survivable nuclear forces as well as more credible employment options. It could turn out that many options for bolstering deterrence in these areas could well complicate future arms control efforts. There is no obvious solution to this dilemma. Some observers, like Christoph Bertram, have argued that the emphasis in new arms control efforts should be shifted from limiting numbers and types of weapons to exploring agreements that would prohibit such 'destabilising' strategic behaviour, such as threatening each other's retaliatory capabilities.[4] Yet asking too much from SALT not only runs the risk of raising expectations that will surely be disappointed later, but it places a national security burden on the negotiatings that they cannot bear. If there is a solution, it probably lies in asking arms control to do less instead of more. Regardless of whether the SALT II treaty is ratified, the United States, in any follow-on negotiations, should not seek severe quantitative reductions or tighter qualitative constraints. Accords that provide both sides with some flexibility for dealing unilaterally with their separately perceived military problems are not only more negotiable but they are also probably more likely to strengthen deterrence. Scaling down ambitions for arms control will not totally eliminate the very real tensions between efforts to achieve agreements at SALT and unilateral efforts to bolster deterrence. But the first gesture towards wisdom in this area is to recognise that such a tension exists; the Reagan priority of improving strategic capability before entering into major strategic arms negotiations is a step in this direction.

While negotiations alone are unlikely to produce solutions to military problems, arms control, in conjunction with American defence initiatives, can offer some promising approaches to coping with new Soviet challenges. Probably the strongest military argument that can be made for SALT II is that by putting a ceiling on Soviet missile warheads until 1985, the agreement could, if extended, work to enhance the survivability of multiple-aim-point basing modes for the MX. However, the main way that defence programmes and arms control can interact in a positive fashion is by enhancing American negotiating leverage to achieve specific objectives.

At the same time, problems can also arise from the inappropriate use of 'bargaining chips' in negotiations. This is perhaps most clearly illustrated in NATO's decision to proceed with the modernisation of long-range theatre nuclear forces. Because of the sensitivity of nuclear deployment issues in some European states, governments are attracted to the idea of coupling an Alliance decision to proceed with the deployment of a new, long-range system with an arms control offer to the East. While this strategy has allowed the Alliance to overcome domestic opposition in West Germany, and less successfully in Belgium and the Netherlands, to modernising American nuclear forces in central Europe, in the longer run, it could backfire. It is because it is hard to see why, at this stage, Moscow would be interested in any arms control proposal aimed at limiting long-range theatre forces. Just as the MX, with an IOC of 1986, is unlikely to arrest Moscow's current deployment of the SS-18, an extended-range Pershing, with an IOC in the early 1980s, is unlikely to coerce Moscow into dismantling its existing SS-20s. Even if the Soviet Union were interested in seriously exploring limits on long-range systems, it has strong political incentives to refuse any negotiations so as to create new doubts in the West over the advisability of proceeding with negotiations. By making arms control, as the Alliance has done, an indispensable component of any programme for restoring a robust, in-theatre nuclear posture, the West has given Moscow an open invitation to interfere in NATO decision-making. At most, if the Alliance proceeds with the deployment of several hundred medium-range systems during the next few years, it is conceivable that Moscow would accept some upper limits on the size of its long-range theatre forces. But even this outcome may not be possible. By hoping (and in some cases, pretending) that there is a negotiating solution to the problem of Soviet theatre nuclear expansion, when in all reality there is not, the Alliance runs the risk of being criticised for using arms control as a 'fig-leaf' to justify new military programmes.

While it has become conventional wisdom that arms control considerations should be taken into account in shaping defence policy and programmes, it is interesting that so little thought is given to how existing negotiations should be adapted to changing military realities. For example, changes might be profitably made to existing approaches at MBFR. While manpower ceilings remain a useful political, if somewhat irrelevant, military goal, ceilings on equipment raise awkward problems in a period in which the Alliance is engaged in upgrading its conventional capabilities and thinking about nuclear modernisation. Given the character of the Warsaw Pact threat, arms control outcomes that constrain the use, rather than the size, of military forces in central Europe would be of some utility. This, of course, is the essential function of 'confidence building measures' – 'associated measures' in MBFR parlance.

But MBFR may not be the best place to negotiate arms control in Europe. Especially if equipment limits, instead of manpower or confidence-building measures, are going to be the focus for negotiations, the French proposal for an arms control forum reaching 'from the Atlantic to the Urals' makes far more strategic sense than continuing to do business in Vienna.

The possibility of revising existing arms control arrangements to bring them more into line with emerging military realities should also not be overlooked. Revision of the ABM treaty to facilitate the deployment of hard-site missile defences is an especially interesting option. So far, it has only begun to have serious consideration, out of a fear that tampering with the 1972 treaty could lead to an unravelling of the entire agreement. The deployment of hard-site defences might not only enhance the survivability of multiple-launch point ICBM basing systems, but it might even rule out the need for deceptive basing modes altogether, by giving a new lease on life to fixed silos.

Finally, while some negotiations might be usefully rechannelled and replaced, others should probably be abandoned. Prime candidates are the Indian Ocean talks, the conventional arms transfer negotiations and the effort to limit anti-satellite systems. Not only is the concept of a 'naval balance' in the region analytically unsound, but the goal of the negotiations – to freeze American and Soviet naval deployments in the area – is most likely incompatible with growing US security concerns in the region. At the very time that the region is taking on greater importance to the West, the Iranian revolution, radical currents in Arab politics, and, of course, the Soviet invasion of Afghanistan, have made naval forces the only reliable way for the United States to project power into the area, for the time being. Arms transfers provide the United States and its allies an indirect means of projecting power into South West Asia (and other regions) and constraints on Western flexibility in this area also seem inappropriate at present. Finally, attempts to ban the testing of anti-satellite weapons when the United States has yet to develop a capability for satellite interception is a dubious enterprise, particularly when this is an arena of clear American technological superiority.

Notes

1. This chapter was written while the author was National Security Correspondent of the *New York Times*.
2. Leslie H. Gelb, 'A Glass Half Full', *Foreign Policy*, no. 36 (Fall, 1979).
3. Hedley Bull, *The Control of the Arms Race: Disarmament and Arms Control in the Missile Age* (Weidenfeld and Nicholson, London, 1961), p. 90.
4. See Christoph Bertram, *The Future of Arms Control: Part II: Arms Control and Technological Change: Elements of a New Approach*, Adelphi Papers no. 146 (IISS, London, Summer 1978).

4 NUCLEAR ARMS CONTROL AND EUROPE: THE ENDURING DILEMMA

Robin Ranger

The central paradox of Theatre Nuclear Arms Control (TNAC) is that it is becoming politically more desirable, but technically less possible. It will be tempting for European governments and bodies politic to blame the Reagan Administration for the failure of arms control in Europe. Yet the failure of arms control in Europe is merely symptomatic of the failure of arms control as a whole.

However heretical it may be to admit it, the fact is that, as an Alliance of fifteen members combined for their collective defence, NATO is inherently unsuited for conducting arms control negotiations. NATO is even less well equipped to compete with the Soviets in the propaganda aspects of arms control gamesmanship. For NATO to produce an arms control proposal, particularly in the TNAC field, fifteen national bureaucracies must produce solutions to technically complex – if not insoluble – problems that touch politically sensitive, vital, national security issues. These fifteen national solutions must then be reconciled, with a single determined state able to block any proposal it considers undesirable. Not surprisingly, NATO arms control proposals are almost always too little, too late, politically, whatever their technical merits. (The great exception was the MBFR idea which countered domestic US pressures for unilateral reductions in American NATO forces.) To be fair, it is so difficult for NATO to come up with any arms control offer that it calls to mind Dr Johnson's chauvinistic comment on women preachers: that it is like seeing a dog walk on its hind legs. One is so amazed to see the thing done at all that one does not notice it is done badly.

For NATO, the dilemma of nuclear arms control in Europe is acute. NATO's 1979 Theatre Nuclear Force (TNF) modernisation decision was a two-track one, stressing the Alliance's preference for arms control limitations on TNF to halt the intensive Soviet TNF expansion that made NATO modernisation necessary. But if, as is now undoubtedly likely, TNAC proves unobtainable, NATO TNF modernisation will have to proceed, despite the reservations of some Alliance members, notably the Netherlands and Belgium, and domestic opposition, as with the British Campaign for Nuclear Disarmament. As the modernisation debate has shown, there are crucial differences of perception between Europeans and Americans on the feasibility and negotiability of TNAC. The general European view is that it is desirable politically, must be feasible technically, and can be negotiated with the Soviets. The Americans see TNAC as technically not feasible, and certainly not acceptable to the Soviets, even though, perhaps, politically desirable.

These views are a typical manifestation of the different ways NATO-Europe and the US look at arms control as a whole. The Europeans view arms control as essentially a political activity, the technical details of which are subsidiary and relatively unimportant. The Americans have, in the past, seen arms control as an apolitical exercise. Neither view is wholly correct, and the new school of arms control emerging in the US, the 'revisionist arms controllers', have recognised this, stressing the political, as well as the technical, aspects of arms control. The Europeans have been slower to adjust their thinking on arms control to take account of the real technical problems involved. They have also been reluctant to realise that the Soviets have rejected Western-style arms control, while being simultaneously engaged in a massive military effort which exceeds any conceivable defensive requirement.

Since the Reagan Administration will be dominated by revisionist arms controllers, these intra-Alliance differences on arms control will be sharpened by the broad philosophical differences in approaches to defence issues. The European style in strategic discourse is long on generalities and short on specifics; the American style is the reverse. For Europeans, too, it will be difficult to grasp the full extent of the change in the US political landscape, and hence in US strategic policy, brought about by the 1980 elections. But it bears repeating that the Reagan Administration will only accelerate, not cause, the inevitable demise of nuclear arms control in Europe; theatre nuclear arms control was never a serious possibility.

US scepticism over arms control in general, and arms control in Europe in particular, will thus confront West European optimism about strategic and theatre nuclear arms control. The resultant tensions within NATO will be exacerbated by the Soviet Union, which has rejected all efforts to achieve effective arms limitation, inside or outside Europe, but poses as the champion of military detente and arms control. Their political approach to arms control – political arms control – has been so successful for the Soviets (as in their 1978 propaganda campaign against the Enhanced Radiation Weapon – the neutron bomb), that they will not abandon it in favour of serious theatre nuclear force negotiations. These have lent themselves to the obvious acronym of Theatre

Nuclear Arms Limitation Talks (TALT), suggesting that SALT III would be complemented by TALT talks and, ideally, a TALT I agreement limiting TNF.

Soviet TNF modernisation and expansion, which has destabilised the Eurostrategic balance, has significantly increased with the introduction of 5 new missile systems (SS-21/-22/-23 Short Range Ballistic Missiles – SRBM and SS-N-9/-12 Sea Launched Cruise Missiles – SLCM) and an acceleration of the SS-20 Variable/Intermediate Range Missile (V/IRBM) deployment rate (to one every five days). A Soviet TNF modernisation programme of this size and scope is too big and too important to them to be limited in any significant way by arms control. Not only will NATO TNF modernisation be necessary to preserve the balance of deterrence in Europe; it will have to be increased.

The broader political context in which European arms control must be considered will also be extremely unfavourable, if not fatal, to arms control, and inimical to detente. SALT II, the symbol of both, will be effectively scrapped, although officially buried by being melded with SALT III. If NATO-Europe wishes to preserve detente, or, more specifically, the intra-European and intra-German detentes, it will have to separate these, as far as possible, from arms control. This will not be easy, since the Reagan Administration also espouses the linkage of arms control to broader political issues.

It is no coincidence that the European faith in arms control as a substitute for defence policy has lasted longer amongst those who have had less exposure to technical obstacles to arms control, as well as to the Soviet rejection, in practice but not in propaganda, of Western-style technical arms control. No serious student of the SALT negotiations should have had much hope for SALT or TALT producing anything more than symbolic, politically useful arms control devoid of any technical effects. But precisely because the West European public debate on these issues operates at a very generalised level, and is characterised by a cheerful disregard for the basic technical issues, as well as less relevant technological minutiae, this lag in American and European perceptions of arms control will continue to cause major frictions in the Alliance. Neither these frictions, nor their causes, are particularly new, but the implications are far more serious than in the past.

In a very broad, but valid, sense, the new debate on arms control in Europe looks more and more like that of the late 1950s, triggered by the 1957 NATO decision to deploy two-key US nuclear weapons for the defence of NATO-Europe. Then, as now, the US and the major NATO-Europe governments agreed that these new weapons were needed for their defence in the face of a Soviet conventional and nuclear force build-up, coupled with uncertainties over their political intentions. Then, as now, NATO measures to preserve the balance of deterrence on which West European independence depended was opposed by European politicians and intellectuals who felt security could be better achieved by disengagement and arms control. This view was encouraged by a Soviet Union which none the less rejected all such proposals.

Inevitably, now, as then, the result will be the creation of a false dichotomy between the needs of defence and the needs of arms control, tempting the Europeans to blame the Americans for missing a real chance for arms control and detente. Chancellor Schmidt's continuing calls for SALT II's ratification are setting the scene for this to happen. It encourages the fallacious accusation by other Europeans that the US rejection of SALT II represents the rejection of a technically, as well as politically, effective arms control agreement.

In the same way, Europeans are setting themselves up for a second disappointment. They are advocating that President Reagan, instead of adhering to the Democratic party's 'false' image of 'Reagan-the-war-monger', need not really change anything much in US defence and foreign policy. This is a dangerously misleading assumption, perhaps reflecting hope more than sober analysis. The new US Senate will not pass SALT II even if President Reagan asked it to, and he will not do so. On the contrary, his Administration will re-examine the basic assumptions underlying SALT, to ensure that any future SALT agreements take account of the US need for survivable strategic nuclear forces adequate for *extended* defence of attacks on the US and her allies. West Europeans have a chronic habit of forgetting that their security depends not on strategic parity between the Superpowers, but on a US guarantee to go to strategic nuclear war on their behalf. Europeans also suffer from an inability, even after all these years, to grasp that the numbers really *do* matter in nuclear deterrence and that, *pace* McGeorge Bundy's over-quoted misinterpretation of the evidence, one bomb on one city to one hundred bombs on one hundred cities is not an adequate deterrent. This, of course, makes foolish the European hope that President Reagan and the Republican Senate should not change the course of US policy. The new US President, the new Senate and, all importantly, their key staff advisors plus their civil-service appointees, are 'hardliners'. They are also much more conservative, by conviction, than any government West Europe and Canada have seen for the last twenty years or so, save for the Conservative Thatcher government in Britain. Whether one agrees with their *Weltanschauung* or not, it is no use pretending that it is not very different from that which the Europeans have become used to in Washington.

All of the above is a necessary introduction to the central point: European arms control has *never* offered a way out of the dilemmas posed by TNF modernisation. TALT has never had a chance of producing TNAC, because TALT has been based on the SALT model for arms control, and SALT has failed. SALT's failure, in turn, is part of the general failure of arms control to fulfil great and unrealistic expectations. Major NATO-European governments, notably those of Britain and France, have already come to that conclusion. Some of the smaller NATO-European governments, have not. Perhaps more importantly, the general public and the media in Europe has had these hopes of arms control maintained by politicians and analysts advocating arms control measures that they knew, or should have known, were impossible both technically and politically.

The central problem of arms control is that it has been asked to do too much. Amongst the general public in the West, the belief has grown up, in the last decade, that arms control can, and will, provide more security at less cost, with fewer weapons. Arms control has thus tended to become a *de facto* substitute for defence policy and defence spending, and a means of avoiding politically awkward decisions. The two crucial assumptions underlying what can now be seen as excessive hopes for arms control were that, technically, arms control measures, unilateral, bilateral or multilateral, could be constructed to achieve these objectives, and that, politically, the will existed to put such measures into place, particularly between the Superpowers. Both assumptions supported one another synergistically: if arms control could work, technically, there would be more political drive behind it; if there were sufficient political will to make arms control work, technical difficulties would be overcome. The temptation, not always resisted by arms control enthusiasts, was to make the wish that arms control *could* succeed father to the thought that it *would* succeed. This was possible in the 1970s because arms control, like all new forms of salvation, had the advantage of not having been tried and found wanting. But by the 1980s, arms control in general, and nuclear arms control in particular, *had* been tried, and found wanting.

Progress in arms control is impossible unless the central paradox of arms control can be resolved more successfully: that arms control is often impossible, where necessary, and unnecessary, where possible. Instead of arms control arrangements offering a technical end-run around political antagonisms, the technical obstacles to constructing any but the simplest agreements have increased. But if arms control agreements *are* simple, they must involve a great deal of mutual trust that the spirit, as well as the letter, will be observed. They will also have little or no technical substance. The only arms control measures that are likely to work will thus be those that are largely, if not wholly, symbolic. Otherwise, the technical obstacles to concluding and implementing arms control agreements will create suspicions of political bad faith, and political disagreements will affect the already very difficult search for ways in which to overcome these technical obstacles. These were the reasons SALT II took so long to negotiate, and would not have received Senate ratification.

For theatre nuclear arms control, and for NATO, these general propositions about arms control are extremely important politically. They imply that expectations within the Alliance about the possibilities for TNAC will have to be re-examined, and downgraded. Past experience with arms control shows that there is absolutely no chance of slowing up the Soviet TNF build-up which necessitated NATO's TNF modernisation, and which will continue unabated for several years. The Soviets will *not* accept any significant limitations on their TNF build-up being prepared to do no more than talk about TNAC, or, more accurately, to go through the political and institutional motions.

The onus of proof is now really on those who would argue that theatre arms control is feasible and negotiable. The sorts of limits usually discussed, on numbers of Soviet Backfire bombers and SS-20 IRBMs, and of NATO GLCMs and Pershing II IRBMs, would produce instability. They would leave too many other systems unlimited, like Soviet SRBM and SLCM, and could not be verified with sufficient accuracy. This, of course, is over and above the question of whether or not the Soviets are prepared to accept real limitations even in these restricted areas.

Put simply, if SALT cannot work, either as a communication process, or as a means of producing effective arms control agreements, TALT cannot work. Until this year (1981), most discussions of theatre arms control, certainly outside analytical circles, assumed SALT as a model for successful arms control negotiations and agreements. TALT was thus an idea whose time had come. The two major sets of difficulties were the much greater technical obstacles to constructing a TNAC regime, as compared to a SALT regime, and the political problems in bringing the relevant NATO-Europeans (Britain, France and West Germany) into the negotiating process. There were also the separate, but related, problems posed by the British and French nuclear forces, strategic, theatre and tactical (battlefield). Reservations came explicitly from Paris, which rejected in principle the idea of any limitations in SALT (or TALT) on French nuclear forces. Implicitly, London suggested that British nuclear forces, too, could not be limited, nominally because of the practical difficulties of doing so. In the US, arms control advocates extremely influential in the Carter Administration saw TALT I as necessary, possible and desirable. Their critics – some, but by no means all, associated with the Republican Party – saw a theatre nuclear force agreement as undesirable, since unworkable, but probably unavoidable.

These assumptions in favour of a TALT I modelled on SALT I and II, must now be reversed. Significant sections of SALT II, including those bearing on TNF, will have to be renegotiated, notably the Protocol, which only runs until 1983. The SALT model for TNAC is thus no longer valid in two important senses. First, it can no longer be seen as an example of the way in which the political requirements for arms control agreements can override the technical objections. Secondly, the SALT criteria for defining central, strategic nuclear systems, and the means of limiting these, are no longer adequate for SALT, or for TALT.

The pervasive scepticism of arms control in the US strategic and political communities has become so widespread that the general approach has been one of 'back to the drawing board' for arms control and SALT. As a process, SALT might continue if it has to, politically, but is not expected to produce any technically effective limitations. (If SALT II was SALT I.1, SALT III would be SALT 0.0202.) Similarly, TALT – inside or outside SALT III – might be useful as a process, but should not be expected to produce any real limitations on TNF, and certainly nothing like enough to justify halting or slowing NATO's TNF modernisation.

Scrapping SALT – certainly scrapping SALT II – would destroy much of the rationale of the arms control bureaucracies in the US, and her allies. The SALT process could survive *if* it were useful – it is not clear that it is – after being reconstructed on terms equitable for the US and her allies. Similarly, abolishing, or purging, ACDA would help turn the US bureaucracy around in the direction revisionist arms controllers want. Under President Reagan and a Republican Senate, these directions will be pursued much further and much faster.

In the long run the major NATO-European governments might welcome such a drastic reconstruction of the US approach to arms control. The British and French would certainly welcome any US moves to gut SALT and TALT of any technical substance. Such an apparently controversial suggestion makes sense if the interests of NATO-Europe states are sought in their respective governments actual, as opposed to declaratory views. France has always opposed SALT as an instrument of Superpower condominium, and of US domination of NATO. Logically, the French are correct in arguing that it would be absurd for them to agree to limit French nuclear forces as a result of US-USSR SALT/TALT negotiations, and in doubting that a balanced theatre agreement is possible. France has pushed the concept of a Conference on Disarmament in Europe (CDE), since 1978, but as a political rather than an arms control venture. British defence interests are essentially the same as those of France – deterring a Soviet invasion of, or nuclear strikes against their homeland, and of, or against, West Europe. Britain, like France, is modernising her nuclear forces, notably by purchasing the Trident SLBM from the US, a move that will also substantially increase her strategic/theatre nuclear forces. No British government can be interested in limits on British nuclear forces except as part of far more effective SALT/TALT negotiations than are now possible.

West Germany, on the other hand, has a unique interest in TNAC. As Chancellor Schmidt has stressed, she is particularly threatened by the failure of SALT to limit the growth in Soviet TNF. So unlike Britain and France, West Germany has more to gain than lose through TALT talks. Even if these proved as unsuccessful as the MBFR negotiations, the Federal Republic would have tried, and *been seen* to have tried for arms limitation, but if this cannot be achieved, as is likely to be the case, TALT could expose West Germany to even more intense Soviet propaganda attacks. Defusing TALT could thus be in West Germany's interest, *if* this did not sacrifice real chances for arms control. Schmidt's July 1980, visit to Moscow reflected these pressures to seek theatre arms control through TALT, and to do so seriously, but to limit the damage, perhaps postponing TALT, if this appeared to be unobtainable.

In addition, Britain and West Germany both worked strongly for NATO's TNF modernisation decision, a decision that paralleled France's unilateral TNF modernisation. These two major NATO-European powers therefore opposed the smaller NATO members who argued that NATO TNF modernisation foreclosed the chances of arms control, and so should be deferred. Holland has been the most vocal advocate of postponing TNF modernisation, but her attitude was shared, in varying degrees, by Belgium, Denmark and Norway. All four countries have also argued, officially or unofficially, that TNAC is possible, once SALT II is ratified, as they thought it would be. Scrapping SALT will make it harder for them to argue that TNAC can work, although they might, temporarily press harder for the success of institutionalised talks.

The NATO-European attitude to SALT, TALT and TNAC is thus much less favourable than official government statements and public discussion suggest. Politically, all NATO-European governments have to be seen to be for arms control and detente, at least publicly. Privately, Britain and France are opposed to TNAC because it would limit their nuclear forces, vital for their defence, and are privately opposed to even negotiations, because these will give rise to unrealistic expectations for the resulting agreements. For similar reasons, Britain opposes the French CDE proposal: it will encourage false hopes for arms control. France argues it will provide a venue for discussing European security issues in terms broader than those of arms control and give Europe a much needed input into arms control discussions. West Germany is, perhaps inevitably, ambivalent, because she remains in the front-line geographically, militarily and politically: the only NATO Great Power without nuclear weapons, but providing bases for US and NATO nuclear weapons and delivery systems. Her interests would be served by TNAC, if obtainable, or if TNAC were demonstrably not obtainable, it might be better to downplay those talks which do take place. West Germany has the most to lose if SALT is scrapped, in the sense that it is, perhaps wrongly, seen as essential to a detente in which Germany has a unique interest because it provides the framework for the intra-German detente. Yet a new US approach to SALT could, and should, accommodate West German arguments for the need to consider more properly the theatre nuclear, as well as the strategic nuclear, balance.

The trend in the US and NATO-Europe is thus to reassess the chances of TNAC, and hence the feasibility of TALT in the light of the reassessment of SALT and of arms control as a whole. Under these circumstances, talks about TALT talks on TNAC could keep alive false hopes. If, of course, useful measures no longer seem possible, some of the difficulties NATO has encountered in its modest TNF modernisation could be lessened. NATO could argue that its two-track approach – arms control first, TNF modernisation second – has been vindicated. Arms control had proved impossible, so TNF modernisation is unfortunately necessary.

Overall, the Alliance attitude towards arms control, particularly to theatre nuclear arms control, seems like that of the Hippocratic Oath towards the terminally ill . . . 'Thou shalt not kill, but thou shalt not strive officiously to keep alive.'

Obstacles to Strategic and Theatre Nuclear Arms Control in the 1980s

Such attitudes towards TNAC become much more tenable once the very real practical difficulties are understood. Indeed, a major reason for the difference between NATO-European governments' views of TNF modernisation (in favour) and TNAC (sceptical), and the views of those outside of government (usually against TNF modernisation and optimistic on TNAC), is that there is little European discussion of technical issues in the public domain. The contrast with the US is obvious, but the effects are really much more pervasive than is often realised, especially in terms of the media. This point needs stressing, in the TNF/TNAC context, without falling into the analyst's common error of assuming that once the facts are known, the appropriate policy conclusion will be – or even can be – drawn. Serious critics of TNF modernisation, such as the Dutch MP Klaus de Vries, are as knowledgeable as anyone about the issues involved. Yet, after acknowledging the scope of the Soviet TNF growth, and the threat this creates to NATO forces and the strategy of flexible response, they will often make a jump in logic to conclude that NATO TNF modernisation is premature, and that, contrary to the evidence of the last 22 years of arms control, the Soviets would negotiate seriously. The high level of technical illiteracy in NATO-Europe certainly makes it much more difficult for NATO to argue a case for TNF modernisation that depends on understanding of the basic technicalities. So also for arms control, especially as any suggestion that there are real difficulties invites the accusation that chances are being discounted because the analyst does not want TNAC to succeed. Without underestimating the sensitivity of these issues, it is also fair to say that NATO, as a collectivity, is excessively reluctant to discuss *any* issues involving TNF, TNAC and its declaratory strategy of flexible response. Understandably, the argument is that any evidence is more likely than not to be used against Alliance policies. Against this, the Alliance case too often goes, by default, to its critics.

Hence, in part, the problems with TNF modernisation. The case for a positive decision was not understood in all member governments, and certainly not by much of their electorates. Similarly, there is insufficient understanding of the problems in securing real theatre arms control regimes, and hence of NATO making some intelligible arms control offers on TNF. One result was that Chancellor Schmidt publicly raised the idea of a moratorium on TNF modernisation several times, contrary to NATO policy, discussing it in his July 1980 Moscow visit.

His motives were, apparently, to try to secure a freeze on Soviet TNF modernisation, before this gave them total long-range TNF domination, and to convince the West German electorate that he had tried to get an arms control alternative to modernising TNF. On both counts, his approach was a great deal more effective than NATO's continuing failure to produce an arms control offer, or to explain why it had not done so, or when an offer could be expected. Schmidt secured some nominal Soviet concessions, but no movement towards his moratorium proposal. He had, however, been seen to do something for arms control. In this sense, he had taken a leaf out of the Soviet manual of political arms control, or – as it used to be called – arms control gamesmanship. There must also be a strong supposition that Schmidt had looked at the technical blocks on the road to TNAC, and concluded that they were impassable. The following examination suggests that, if he did so, he was correct.

If nothing else, SALT suggested that agreement could be reached, *for the purposes of arms control negotiations*, on the categorisation of weapon systems and their characteristics. Central, strategic nuclear systems had been defined on the basis of range and payloads, and limits imposed on the number of their launchers or launch platforms (SSBNs and strategic bombers), and their warheads (MIRVed and non-MIRVed). However questionable these definitions might be on technical grounds, they had apparently worked. So, too, it seemed, could a similar rough-cut approach to TNF definitions and limitations. Unfortunately, technological changes made it doubtful that even the SALT categories could be agreed on for much longer, much less that new, and much more difficult definitions be established for TALT. Hence the suggestion that new approaches to arms control categories and limitations be established, perhaps on the basis of strategic missions. Even this approach seems unlikely to work, in the sense of forming the basis for effective restraints on TNF in terms of numbers and capabilities, since the same systems can serve different missions, and there is little consensus on what systems are designed to serve which missions on the other side.

These difficulties in agreeing on TNF characteristics are, moreover, only symptoms. The real cause of the failure to agree on definitions, categories and characteristics is the divergence of strategic doctrine and political interests between East and West. Nuclear forces are asymmetrical because they are designed to protect different interests by performing different missions. On the whole, in the last decade US and NATO forces – nuclear and conventional – have been configured for deterrence, both to prevent war, and to limit and terminate a Soviet invasion of West Europe should it occur. NATO's strategy of flexible response accepts that NATO's conventional forces cannot sustain a conventional defence for long – how long is a crucial and contentious issue – before using TNF and threatening to escalate to strategic nuclear exchanges. The idea of war-winning is absent from NATO strategy, as it has been (perhaps until recently) from US nuclear strategy. The Soviets, in complete contrast, hold to a doctrine of deterrence through war-winning capabilities. Their force characteristics suggest that the Soviets have never accepted the two ideas basic to US and NATO strategy, that nuclear deterrence will last almost indefinitely, and that very low levels of Assured Destruction could make deterrence work. For NATO, flexible response often seems to mean, particularly to NATO-Europe, deterrence through uncertainty: if NATO does not really know when, and how, it will use nuclear weapons, the Soviets cannot know, and cannot take the risks involved in finding out by initiating a full-scale invasion of West Europe.

The Soviets are more cautious, to the extent that their concepts of deterrence and security involve so much over-insurance that it guarantees the insecurity of the US, NATO-Europe and China. Their approach to deterrence seems to be that the Soviet Union can only deter an attack on her homeland by being able to defeat the attacker(s). Soviet forces and Soviet defences must limit damage to Soviet territory – if necessary, by pre-emptive strikes – following which Soviet forces must punish, or occupy, the aggressor(s). (This means that Soviet forces must be superior to all possible combinations of opponents, making them insecure as a result, and producing the very encirclement of which the Soviets continuously complain.) They are also sceptical that their massive forces will prevent war from occurring, which thereby increases their interests in war-winning and damage limitations. The influence of the Great Patriotic War is obvious, but the overall Soviet Communist, let alone Russian, experience has not been such as to encourage faith in the rationality of deterrence.

The Soviet Communist Party also has a security problem that the US and most of NATO-Europe lacks: not China, but the internal legitimacy of their rule. In Eastern Europe too, this is a major source of insecurity with the local Communist Parties depending for power on Soviet occupation forces. In this connection, Western Europe poses an involuntary threat to Soviet security which the Soviet leadership would presumably wish to eliminate *if* the opportunity occurred, or if the threat became intolerable. The US and NATO-Europe therefore have to deter this threat. Such an asymmetry of interests has produced remarkably asymmetrical strategic and tactical nuclear forces, with such different characteristics as to make meaningful TNF limitations impossible.

The basic characteristics of TNF make it very difficult to define different types of weapons in the clear-cut, unambiguous way required for arms control negotiations and agreements. A broad distinction can be drawn, on the basis of range, between long, medium and short-range TNF, with Battlefield Nuclear Forces (BNF) forming a separate category. Yet even this distinction blurs at the boundaries, especially with the new Soviet systems deployed in the last few years. Beyond this, no real categorisation of TNF is possible, except in terms of broad classes of systems. Otherwise, the chief characteristics used as the basis for SALT I and II limitations are all subject to such wide variations as to be meaningless. Not only that, but the different classes of systems are non-comparable; NATO and the Warsaw Pact have quite different force-mixes and, as noted above, their missions are different. Negotiations for TALT I could face an impossible task in trying to balance, for example, Soviet short-range SLBMs against NATO GLCMs.

LRCMs represent a special kind of problem. Although usually described as gray-area weapons systems, they are really more representative of what could be called Variable Range Systems/Launch Platforms. Their range can vary from 2,500 miles to, say, 250 miles, according to mission and payload, covering the range spectrum from strategic to short range (Theatre). In addition, A/SLCM are carried on Variable Range Launch Platforms: aircraft, ships and submarines. Soviet TNF modernisation has, not surprisingly, included longer range SLCMs, also likely to be deployed in the GLCM mode. Their SLCMs are deployed in large numbers on Variable Range Launch Platforms, including dedicated conventional and nuclear submarine launchers (SSC/SSCN). At present, Soviet ALCM have more limited ranges than Western LR-ALCM, although, logically, they may be expected to deploy longer range ALCM. Carrier based aircraft are also Variable Range Systems with Variable Range Launch Platforms. At present, only the US and French carriers could count as LRTNF, but the Soviets may soon deploy their first fleet carrier, and there are also smaller British and Soviet V/STOL aircraft carriers. Some idea of the problems this might create, in definitional terms, is given by considering the case of US ALCM, on an A-7E strike aircraft. How could the range be calculated for an ALCM (maximum range 2,500 miles) on an A-7E (maximum range 2,800 miles) on an aircraft carrier (which can travel, in one day, over 700 miles)? Hence the opposition to US/NATO deployment by arms control advocates.

The SALT II solution to the LRCM aspect of this problem shows its seriousness. Heavy bombers were arbitrarily defined so as to exclude Backfire; only heavy bombers were allowed to carry any LR-ALCM, and those doing so had to be identified by Functionally Related Externally Observable Differences (FROD). However ingenious, the solution was unsatisfactory, temporary and politically very difficult to sell. This emphasised that not only are the analytical problems of defining TNF for the purposes of arms limitations substantial, but that any solutions must be acceptable politically, and to public opinion, in the NATO countries. Analytically, it may – or may not – have been defensible to exclude Backfire from SALT II, but it was politically indefensible. Backfire looked like a strategic bomber to US Senators and electors, so excluding it, while trying to limit its performance and production via a letter from Mr Brezhnev, *looked* like appallingly naïve negotiating. Yet no other solution was available, since the Soviets would not agree to count Backfire, with ALCM, as strategic weapons.

The SALT II Protocol ban on G/SLCM deployment demonstrates the increasing incompatability of arms control and defence requirements. Banning G/SLCM deployment was really the only effective – or negotiable – limitation possible, so the Protocol was clearly a precedent to be extended in SALT III. Administration assurances to the contrary were simply not seen as credible, given its commitment to arms control, even after NATO's decision to deploy GLCM. The difficulties in drawing even such a basic distinction as that between central and theatre systems were thus threatening SALT. This makes it difficult to see how the much more complex distinctions needed for TALT could be drawn.

The Soviets, for their part, have a number of strongly held positions which would make it impossible to resolve those technical obstacles to securing a meaningful TNAC regime. These positions seem to be more than merely bargaining chips, although they are naturally used as such. For example, the Soviets say that TNF limitations must include *all* FBS capable of hitting Soviet territory. They have always argued, in SALT, that such systems are really strategic, and were still arguing this in the last round of SALT. Such a definition could give them a handle on the British and French nuclear forces and let them count these, as well as *all* NATO long and medium range TNF, against Soviet LRTNF. The latter would presumably be defined as only those M/IRBM and Backfire bombers normally based in the Western USSR.

Interestingly, the Soviet definition of what constitutes LRTNF for arms control purposes is really that of Eurostrategic forces. As such, it is essentially the one preferred by NATO-Europe for Soviet TNF, that is, that all Soviet TNF capable of reaching NATO-European homelands are strategic. For West Germany, this includes Soviet Battlefield Nuclear Forces, but even for Britain and France, would include Soviet long and medium range TNF, including cruise missiles. NATO-Europe would clearly insist that, at a minimum, SALT III reject the SALT II concept that Soviet deployment of gray-area systems like Backfire and the SS-20 against NATO-Europe did not matter (encouraging the Soviets to increase their deployment of such systems).

Examining TNF by range also highlights more general asymmetries in Soviet and US/NATO TNF. Soviet TNF are much newer, much more numerous, and of much longer range. Basically, Soviet second-generation (1970s) TNF are being deployed against first-generation (1960s) NATO TNF, a point most NATO governments understand – hence TNF modernisation – but have not properly explained to their publics. In addition, NATO TNF are more focused on battlefield and medium-range operations, another asymmetry based on doctrine and differences in interests and capabilities for escalation dominance.

Overall, it is impossible to see how any adequate definitions of TNF can be arrived at, let alone form the basis for limitations or reductions. This is not an original conclusion; it is, however, one that NATO-Europe will come to realise later than the US. Hence the paradox that NATO-European governments have to appear to be for TALT and TNAC even though they know, or suspect, that TNAC is impossible, or, like Britain and France, would have to oppose TNAC if it *were* possible.

The US has faced the same paradox in a different setting. Politically, the domestic pressure for TNAC came not from the general public, but from arms control supporters, including those in the Carter Administration. They knew that any technical analysis showed that effective TNF limits would become more and more difficult to obtain with TNF modernisation. Hence their (successful) advocacy of delays in modernisation, including cancellation of ERW, and the SALT II limits on LRCM. The Soviet failure to recriprocate these gestures of unilateral arms restraint had, however, made it impossible to delay any longer NATO TNF modernisation. In response, the arms control community resorted to what can only be called a leap of faith: the assertion that TNAC would work, despite the evidence that it could not. Arms control enthusiasts aside, there seemed to be little support for TNAC in the US defence community, except as a political damage-limitation operation to soften the impact of TNF modernisation. This made TNAC an example of the revolution of declining expectations for arms control, although it will stay alive, as an issue, for a couple of years, since the Soviet Union will use it to try to derail NATO TNF modernisation.

Conclusion: Arms Control as History

The Reagan Administration's objective will be to radically restructure the existing arms control concepts forming the basis for US and NATO policies. This may be done by a frontal assault, terminating existing arms control negotiations and institutions, or indirectly, by reconstructing them, i.e. by being for arms control in principle, but against it in practice. The latter approach was used by Republican opponents of SALT II, both genuinely and disingenuously. The former may recommend itself on the principle of introducing unpopular measures in the first years of a new government. A mixed approach is also possible, balancing scrapping SALT, which will upset West Germany and the smaller NATO-European powers, with a propaganda push for TNAC coupled with increased TNF modernisation. Simply put, the Reagan Administration will *not* expect early arms control negotiations to produce arms control agreements which the US could accept.

Effective arms control alternatives to TNF modernisation will not be available. Soviet TNF modernisation will continue at the same rate, forcing NATO to continue its counterbalancing TNF modernisation (and perhaps to increase it). The Alliance will continue to experience the political problems this involves. NATO will, therefore, have to continue to talk about TNAC and will be under more, not less, pressure to be seen to be in favour of TNAC until this becomes obviously impossible. The Soviet Union will increase this pressure by intensive propaganda and symbolic gestures, designed to persuade NATO-Europe and other West European public opinion that the Soviets are in favour of arms control and detente, which are in turn being blocked by NATO. Such propaganda will have considerable appeal if Soviet-American relations continue to deteriorate, as regrettably seems likely. Symbolically, the demise of SALT II will mark the death of the Superpower detente. SALT II will not now be ratified and will be renegotiated by the Reagan Administration, a euphemistic way of saying that SALT I and II will be scrapped, and the SALT process transformed into another means of pursuing the US-Soviet conflict. Therefore, since the SALT talks (SALT II/III) cannot deal with, or offer progress on, TNAC, so the pressures on TALT will be temporarily increased.

The solution to NATO's arms control dilemma – it wants TNAC, but cannot see how to get it – will be to substitute symbolic for substantive TNAC. France's CDE proposal, West German suggestions for a moratorium on Soviet TNF increases, and the general West European pressure for more extensive CBMs, inside or outside the CSCE Final Act framework, are all evidence of support for symbolic arms control measures. They have not, and are not, likely to produce effective arms control in Europe so long as this is rejected by the Soviet Union. But they have enabled NATO-European governments to convince their electorates that they have tried to secure progress in nuclear and conventional arms control, and in improvements in the political aspects of European security. In so far as they have failed, and will fail, they can put the blame on the Soviet Union, rather than on the technical difficulties.

Curiously, perhaps, this is more acceptable politically, despite the widespread European reluctance to see a return to the Cold War. Arguing that theatre arms control is technically impossible may be true, but it sounds, to the public, like a poor excuse. In so far as these symbolic proposals produce continuing talks about effective TNAC, they enable NATO-European governments to argue that they are still exploring the possibilities of arms control, but that this will take so much time that necessary improvements in NATO defences cannot be held up until progress in arms control is obtained. These arguments are genuine. NATO-European governments, especially West Germany's would infinitely prefer effective arms control alternatives to continued Soviet-Warsaw Pact force increases which NATO will have to counter at significant economic and political costs. Realistically, though, they were already sceptical that such alternatives existed, or were negotiable, even before the US elections ruled these unlikely.

The French CDE proposal may thus be an idea whose time has come. Despite all the valid objections that have been made to the idea of a CDE, even its critics and opponents agree that it is attractive to governments anxious to be seen to be working for arms control and detente – especially as the chances of both are rapidly decreasing to zero. In principle, the idea of an all-European Conference on Disarmament is immensely appealing. In practice, it could enable NATO-European governments to argue their own individual concerns in these areas, and criticise the US as well as the USSR.

An obvious analogy would be the early days of what is now the Conference (Committee) on Disarmament (CD). Between 1960 and 1964 the Ten (after 1961, Eighteen) Nation UN Disarmament Conference (T/ENDC) enabled Western governments to demonstrate their commitment to arms control to their electorates and to counter Soviet propaganda successes with one of the most impracticable arms control proposals ever advanced – General and Complete Disarmament (GCD). The T/ENDC meetings also acted as a useful forum for exchanging views and, when the US and USSR wanted arms control agreements to symbolise their post-Cuban Missile Crisis detente, to help negotiations. In some ways, the MBFR talks serve these functions, but are really too limited in terms of membership, forces discussed and geographic areas. The CD has become too big, too diverse and produced too little, to mean much to NATO-Europe or other Western publics; there is also a signal perceptual advantage in beginning *new* negotiations with a *new* mandate in a *new* forum.

A CDE would also give NATO members – including the US and Canada – more scope for exploring their own solutions to the TNAC paradox. From the collective NATO viewpoint, the ideas of individual NATO members, especially NATO-European members, trying unilateral arms control initiatives is appealing, technically and politically, but they would also not make any progress, except on paper. Serious proposals could at least be made and, if the Soviet Union rejected them, would still have improved the proposer's political position at home.

This is not to say that NATO should not continue serious arms control efforts and should, indeed, try to give these a higher profile. But, because of the limits to what NATO can do collectively, especially to counter the stronger Soviet arms control propaganda, the Alliance is unlikely to lose much by national arms control initiatives. These will certainly continue, since NATO members will want to try for what few chances of effective TNAC there are, and to protect their domestic flanks. After all, the US has been the leader in unilateral arms control initiatives, as in SALT, where NATO has been informed, not consulted, as in the ERW cancellation. These pressures for individual initiatives will be acute in coming years, as the chances of arms control and detente are seen to be going down the drain. Thereafter, the situation is likely to be so bad politically that pressures for arms control, other than unilateral disarmament, may cease. Arms control loses its appeal if it is manifestly impossible.

The only systems to which SALT-type constraints might be applied would be fixed-silo I/MRBMs, plus SS-12 and Pershing II SRBMs. Intuitively, since modernised NATO TNF (Pershing II and GLCM) have been justified as a response to a Soviet TNF build-up popularly perceived as consisting of the Backfire bomber and SS-20 V/IRBM, these will probably form the main candidates for TNAC proposals. Comparable systems would also have to be included: for NATO, Pershing I and US FB-III, for the Soviets, SS-4 and SS-5 I/MRBM, SS-12 and SS-22 SRBM, Tu-16 and Tu-22 bombers. The Soviets would insist similar British and French systems be included, though these two countries would reject such demands.

Most interestingly, any proposals for equal aggregate limits *à la* SALT will favour NATO. The Alliance would, in fact, be able to use the basic Soviet bargaining position in the SALT I negotiations: they have to catch up in terms of quality, and quantity, and so actually need quantitative superiority for essential equivalence, given their inferiority in terms of quality. Intelligent Western *proposals* for TNAC would legitimise NATO modernisation.

If these are also the best bases available for effective TNAC, then they reinforce the previous arguments that this is impossible. Any TNF limits that could be devised in theory could be nullified by additional deployments of TNF not limited, and by the transfer of European TNF from other theatres in a crisis. Verification of TNF limitation would also be impossible without adequate National Technical Means (NTM). The objective of any technically effective TNAC – as distinct from TNAC for political purposes – is to enhance the stability of the European military balance. TNAC is *not* an end in itself, so it is not sufficient to find categories of TNF that could be subject to arms control limitations. The question is not whether some TNF could be limited, but whether such limitations would enhance stability. The answer is that no foreseeable, or negotiable, arms control limits on TNF could contribute to stability in Europe. On the contrary, they could create additional instabilities, since additional deployments of non-limited TNF would create justified charges of bad faith, as would the major verification problems involved.

The paradox of Theatre Nuclear Arms Control cannot be resolved, except by time. Within the next few years, almost all of the arms control assumptions NATO has come to live with since 1963, will be changed. The political environment will become so hostile to arms control that the idea of TNAC will join the disengagement schemes of the 1950s as an appealing, but self-evidently unavailable, alternative to TNF modernisation. NATO's two-track approach to TNAC and TNF modernisation will be vindicated not by the success of its efforts to seek TNAC, or appear to do so, but because TNF modernisation will become part of the general modernisation of US strategic nuclear forces, and of US and NATO conventional forces.

Research Note

This chapter is based on discussions held in Europe and North America in 1978-80. Although many of those who gave of their time must remain anonymous, the author would like to thank, in particular, Uwe Nerlich of the Stiftung Wissenschaft und Politik (Ebenhausen) and George Lindsey, Chief of the Operational Research and Analysis Establishment (Ottawa). For their support and institutional hospitality, special thanks go to the Canadian Department of National Defence, NATO, the Centre for Canadian Studies at Johns Hopkins University, and the Centre for International Studies at the London School of Economics.

Those sceptical of the author's scepticism are directed to his *Arms and Politics: Arms Control in a Changing Political Context* (W.W. Gage, Toronto, 1979 and 1980), where some of the conclusions presented in this paper are developed at length. For views reflective of current US Administration attitudes to NATO and arms control, see Barry M. Blechman, 'Do Negotiated Arms Limitations Have a Future', *Foreign Affairs*, vol. 59, no. 1, Fall 1980, pp. 102-25; Walter F. Hahn and Robert L. Pfaltzgraff, Jr, *Atlantic Community in Crisis: A Redefinition of the Transatlantic Relationship* (Pergamon Press, New York, 1979); and David S. Yost (ed.), *NATO's Strategic Options: Arms Control and Defense* (Pergamon Press, New York, 1981).

5 ARMS CONTROL AND EUROPEAN SECURITY: SOME BASIC ISSUES

Colin S. Gray

The Crisis of Arms Control

For reasons of convenience and ease of negotiability, particular arms control processes have a distressing tendency to take as their mandated territories significantly artificial worlds, removed from, indeed often indifferent to, the proper scope of defence planning,[1] and hostile to the integrity of strategy. Christoph Bertram was not exaggerating when he claimed that 'arms control has come, in much of the Western discussion, to exist quite separately from national security policies . . .'[2]

It has become fashionable to argue that 'arms control has essentially failed',[3] though much controversy remains over the reasons for failure. The 'great SALT debate', which never quite occurred, and was terminated as commentators turned to more pressing and serious subjects of defence and foreign policy,[4] showed little sign of shedding light on genuinely fundamental, and fundamentally important, issues. In 1970 one could be sceptical over the promise of formal Soviet-American or NATO-Warsaw Pact arms control processes, but there were acute problems of evidence for all sides of the policy argument. By 1981 many, if not all, of the hopes and fears of the decade past have been resolved by the parade of events.

The demise of SALT II, the *immobilisme* of MBFR, and the unpromising prospects for NATO-Warsaw Pact negotiations on long-range theatre nuclear forces (LRTNF), have combined to have a useful catalytic effect upon Western policy debate. With the diplomatic agenda of arms control in such an inactive condition, former debating adversaries, of all doctrinal persuasions, are beginning to take something of a sabbatical from immediate policy-relevant topics of contention, having been accorded a rare breathing space for more fundamental inquiry.[5] As noted above, the SALT debate in 1979 did not stimulate such inquiry because all debating parties were more concerned with making their own various cases than they were with re-examining the validity of their assumptions. In addition, arms control debaters in the late 1970s were seeking to establish, hold and enlarge a political constituency among the 100 members of the US Senate, and the people believed to be most influential upon those Senators. Such an exercise did not encourage introspection or the raising of basic questions.

The nature and purpose of East-West arms control processes constitute a region of inquiry the outcome of which will be fundamentally important in assessing the merit of particular arms control proposals and agreements. However, in the SALT debate of 1979 the proponents of the Treaty did not wish to risk muddying the waters by inviting rigorous investigation of the character of the SALT process *per se*, beyond somewhat general reference to the near-essentiality of SALT for Superpower detente and scarcely less general reference to the alleged fact that the road to a more perfect arms control future had to pass through the territory of an admittedly imperfect SALT II. The modesty of the seven-year-long SALT II negotiating achievement, although defensible, might have proved to be an embarrassment were debate to have focused *not* on the details of the proposed Treaty, but rather on the Treaty in the context of long-standing arms control aspirations.[6]

Some of the opponents of the SALT II Treaty tended to limit their critique to the details of the Treaty because they sensed that the US body politic was not ready for a frontal assault on the arms control process as a whole. In short, in 1979 one could respectably oppose SALT II but not the SALT process writ large. In many cases, this was a political and not an intellectual judgement. Rightly or wrongly, it was believed by many of the opponents of SALT II that people were not ready to be told that SALT was a fatally flawed exercise. However, it is a fact that many critics of the actual output, or lack of output, of the major East-West arms control processes of the 1970s (SALT and MBFR) are genuinely undecided over the issue of the future promise of arms control. It is worth recalling that a major argument in defence of SALT II was to the effect that this package, negotiated by three Administrations, was the best that could be accomplished. Some critics were unconvinced, insisting that renegotiation of important elements in the agreement was both necessary and feasible.[7] It is not at all obvious that they were correct. Indeed, one is probably on safe ground asserting that SALT II was the Treaty that the United States deserved, having been outspent by the Soviet Union on strategic forces by an average of nearly 300 per cent over the decade 1970-80.[8]

There is a substantial literature critical of the US approach to arms control negotiations in general,[9] and to the SALT exercise in particular,[10] which encourages the belief that better treaties might have been negotiable. Nonetheless, while granting that many errors were committed in negotiating tactics and strategy, strong critics of recent arms control performance need to be reminded of two underlying facts. First, Western approaches to, and performance in, arms control reflect the very character of Western democratic societies: we may be our own worst enemies in the realm of dealing effectively with command-types of political systems in a negotiating process. Yet this cannot be changed without changing what we are as societies. Secondly, even had the United States and NATO-European countries been more skillful in their SALT and MBFR negotiating practices than was in fact the case, the axiom that 'you cannot fight arithmetic' would retain much of its validity. The raw material out of which arms control agreements are fashioned comprises weapon programmes underway, force levels deployed, and technologies to which the negotiating parties have demonstrated a credible degree of commitment to see through to fruition. If there is a major difference in military programme momentum between the parties, that fact will be reflected in substantially unequal agreements. It is generally understood today that arms control agreements can only register facts; they cannot reshape the strategic world in ways uniquely beneficial to Western ideas of what is a stable context.

To date, no commentator has come forward with a politically respectable and intellectually persuasive analysis suggesting that East-West arms control is, on balance, a poor idea. Instead, it is increasingly popular to argue that adequate and useful arms control agreements may be negotiable if the Western Alliance corrects several adverse military trends. Most of the severe critics of SALT II, for example, urged that the Treaty be returned for renegotiation.[11] Although this author can conceive of worthwhile, 'stabilising',[12] SALT and MBFR regimes, he is less than sanguine over their negotiability. On balance, and somewhat reluctantly, this chapter, which approaches the question from both a political and narrower military viewpoint, rests upon the proposition that East-West arms control negotiations tend to have a net negative impact upon Western security.

Arms control has come to have some of the features of an addiction, or a bad habit, for Western political systems. Moreover, this affliction, to the detriment of intra-allied solidarity, has been stronger in NATO-Europe than it has in the United States. For many NATO-European opinion leaders, it would seem, arms control has changed its status from that of an experiment to a permanent, *and essential*, diplomatic institution. Whether that institution encourages the evolution of a more benign military environment is of relatively little importance since

NATO–Europeans, for obvious if perhaps short-sighted geopolitical reasons, tend to view their security more in a political than in a military light.[13] The logic is very simple: there is no acceptable alternative to East-West detente; detente requires the presence of arms control linkage; therefore arms control is essential. This logic summarises what occurred very largely by historical accident ten years ago. MBFR and SALT each had several different roots, but their relationships one to the other and both to the state of East-West political relations, have become clear only in very recent years. Whereas, until the mid-to-late 1970s, it was assumed widely that arms control was essential for detente,[14] more recently it has come to be recognised that, to an even greater extent, detente is essential for the achievement of progress in arms control. In fact, far from being the essential facilitator of detente, arms control, if anything, has come to be a negative interdictor of detente processes.

The so-called 'arms control paradox' has applied: truly substantial arms control measures are negotiable only between countries who do not anticipate having to fight each other. Genuinely constraining arms control regimes cannot withstand the press of negative political events and trends. Since the late 1950s, it has been hoped that arms control technicians could effect an 'end-run' around many political difficulties. It was expected that arms control agreement could be reached on limited, though important, *essentially technical* issues bearing upon the stability of the military environment;[15] that such a limited-focus arms control endeavour might be substantially impervious to the ebb and flow in the climate of general political relations; and that perhaps the practice of agreement and the habit of co-operation might prove to be enduring.[16] In short, arms control could be seen as a learning process wherein both sides would learn to co-operate in narrow areas of clear common interest. This is particularly the case with reference to attempts to alleviate actual or potential 'mechanistic instabilities'[17] that might promote 'the reciprocal fear of surprise attack'[18] in a period of acute crisis.

The 1970s demonstrated that Soviet-American agreement on a subject as central to security considerations as strategic nuclear weaponry requires an exceptionally permissive political climate. On reflection, it is probably accurate to maintain that it was not 1979 that was extraordinary in its political climate (*vis-à-vis* SALT II); rather was it 1972. Although Soviet military intervention in Afghanistan delivered the *coup de grâce* to the prospects for Senate ratification of the Treaty, it was common knowledge that the Administration did not have in hand the two-thirds majority required for ratification even prior to that event – in *early* December.

Here, however, it would be unwise to play down the factor of official ineptitude. Mr Carter sacrificed vitally important months in 1977 on the altar of his well-intentioned, but profoundly ill-judged, goal of achieving major strategic force level reductions in SALT II (the March proposals). The Ford Administration had come very close to securing a defensible SALT II agreement in January 1976, but election year political considerations sufficed to prevent immediate consummation. The months that Mr Carter lost in 1977, both because of his 'back to the drawing board' approach, and because of the change in the political tenor of Soviet-American relations (brought about by his early *démarches* on the subject of human rights in the USSR), comprised time that could not be recaptured at a later date.[19] By 1978, and through most of 1979, Mr Carter's reputation for policy wisdom, and for finesse in policy execution was so low that his authority *vis-à-vis* the key Senators for SALT Treaty ratification was close to negligible.

In terms of the attitude of the US Senate, the prospects for SALT II ratification in 1979 were effectively killed by the delay imposed by the political reactions to the September leak concerning the Soviet combat brigade in Cuba. Mr Carter could have ridden out that storm over so little – certainly nothing that the US could change – had he maintained a low profile on the issue, affirmed prior official US knowledge, and argued the essential triviality of the matter. Instead, he chose to assert that the presence of the brigade was 'unacceptable'. Unfortunately he had no theory of how that presence could be removed. Thus, in fairly short order, the presence of the brigade was found, *de facto*, to be acceptable. The Iranian and then the Afghanistan crises caused Mr Carter's popularity to soar in November and December 1979, but that was not the kind of popularity that could be translated into Senate votes for SALT II. His popular strength early in 1980 stemmed from the fact that he was running against the Soviet Union and against Iranian terrorists. These were domestically popular political enemies, but in the Soviet case, at least, they were not conducive to a US domestic political climate helpful for the ratification of arms control agreements. America apparently beleaguered, is American truculent, particularly in an election year.

This apparent digression on the subject of the politics of SALT is pertinent to the European security theme of this chapter, first, because it is a fact that there can be no 'progress' in European arms control if the SALT process founders; and secondly, because there is an important degree of commonality in the security issues that pertain to SALT, MBFR and the negotiations. In this connection, the following appear to be historically well-demonstrated facts.

Although 'the detente process' of the 1970s engaged many different interests through diverse channels, there can be neither a pan-European, nor a partial-European detente, if Soviet-American relations are deteriorating. Needless to say, the USSR has encouraged just such detente divisibility.[20] In the last resort, Western European security rests upon the transatlantic security guarantee. Equally, in the last resort, it is the United States which determines what are the limits of acceptable NATO-European behaviour *vis-à-vis* the East.[21]

Arms control in Europe probably has no future, just as it has no past. There is an insufficient basis of common interest between the Warsaw Pact and NATO for there to be any serious prospect of an agreement that reasonable observers would deem worthwhile, let alone 'stabilising', by rigorous Western technical definition.[22]

In principle, the only remote prospect for arms control in Europe will flow from the Superpowers agreeing to a radical redirection of their erstwhile strictly bilateral arms control exercise. In terms of strategic theory, NATO cannot seek stability through SALT and MBFR as conducted to date. The basis for NATO's still authoritative concept of flexible response – enshrined in MC 14/3 of 1967 – is a multi-level stability resting upon useful *instability* at the higher levels of potential violence.[23] Stated in the most basic terms, NATO could endure a planned insufficiency of forces in the European theatre, if behind those forces there was a US strategic force posture which could be employed intelligently to redress the theatre imbalance. What one could not do, sensibly, was to negotiate on 'the parity principle' in SALT,[24] continue to be relaxed about the (im)balance within the European theatre, and expect the West's overall security condition to be unimpaired as a consequence.

Discussion of 'Arms Control and European Security' cannot be divorced because of its regional specificity from the more general 'crisis of arms control' that is near-universally acknowledged in the West today. The Soviet invasion of Afghanistan has served to bring to a head a number of very basic questions that have begged for answers for at least the last five years. First, can the West conduct serious nuclear arms control business with a state which has a 'battlefield' (in John Erickson's terminology) or operational approach to nuclear-armed forces?[25] Secondly, are worthwhile measures of arms control negotiable with a state which is obliged, for fundamental reasons of its domestic legitimacy, to define Western democracies as enemies?[26]

Historically, it has always been the case that arms control regimes served to redirect, rather than to halt, arms competition. It is a matter of fact that in the 1970s, with a SALT regime, and ongoing MBFR negotiations, the Soviet Union worked very hard to undermine such 'stability' as existed in the early part of the decade. The Soviet Union appears to see stability at the regional level in terms of the ability to intimidate and to prevail militarily, while at the central level, the Soviet Union, similarly, appears to see stability in an ability to wage a successful military campaign. One can argue that a particular arms control agreement may be of mutual value even if the 'High Contracting Parties' signed for very different reasons. But one should not refrain from observing that Soviet and US/NATO-European 'strategic cultures' are so different,[27] in the context of very porous and very partial arms control agreements, that there may well be an insufficient conceptual basis for fruitful co-operation.

Causes for Concern

This section of the chapter examines a set of historically defensible claims concerning the merits of formal arms control negotiations in general, and the SALT and MBFR processes in particular. 'Historically defensible' means that the claims are generally supportable by references to real, and by and large recent historical events, as opposed to being supportable solely by particular chains of logic. By way of illustration of the latter, one could point to the 1970 vintage theory that US ABM deployment would promote arms-race instability. It so happens that there has been arms-race instability in the *absence* of ABM deployment. Had there been no ABM Treaty, it is very likely that many people would have attributed the Soviet MIRV deployment on the fourth-generation ICBMs (SS-17s, -18s, and -19s) to a determination to overcome US ABM deployment.

1. Arms control institutions (such as SALT and MBFR) tend to become vehicles for the belief that the process of competitive armament is more of a danger than are Soviet capabilities.

The problem in arms control negotiations is not to control the arms race; rather it is to enhance security. As a generalisation, and to simplify without doing undue violence to the evidence, arms competitions very rarely, if ever, run 'out of control', and still less do they 'cause' wars.[28] States compete in arms because they discern a threat and anticipate that they might have to fight.[29] The vision of Soviet, American and NATO-European arms control technicians labouring in the SALT and MBFR vineyards, bent upon achieving the *common* goal of taming the arms race, is pure and dangerous fantasy.

If there should be a war in Europe, the audit trail of proximate responsibility may well lead to revolting East Germans, dissident Croats, or imperially expansive-minded Great Russians. It is most unlikely, however, to lead to tank and anti-tank competition, and the like. The existence of arms control institutions provides a focus, and an opportunity, for misguided (or historically ill-educated) people to ignore the inconvenient facts that their country and alliance is actually engaged in a competition with an adversary who means them no good, and that this competition could have a wide range of politically and militarily meaningful outcomes.

The very fact of an arms control institution tends to alter the political context for the discussion of the weapon systems that fall within its negotiating mandate. MX, or cruise missiles, or an ABM system, assume transient diplomatic importance because they are on today's arms control agenda. Naturally, if unfortunately, individual weapon systems come to be assessed overwhelmingly from an arms control perspective, in terms of the problems of negotiability, verification and the like. Arms control, far from constituting a useful adjunct to defence planning and strategy comes to assume the lead in the provision of criteria for weapon system assessment.[30]

Many people continue to see SALT (and MBFR, *ab extensio*) as 'the last best hope of mankind'. As with any major addiction, such people cannot imagine life without the crutch of arms control. Lest the point be lost, there can be no debate over the worth of the goals of arms control. No responsible defence analyst can be indifferent to the risks of war occurring; the prospective scale of damage should such war occur; or the burden of peacetime defence preparation. The major caveat in need of specification today is that it is not at all obvious that these unexceptional goals of arms control can be pursued usefully through formal arms control negotiations.[31] Moreover, formal arms control negotiating processes tend both to focus public attention upon the military dimension of relations, and to produce a degree of sensitivity to static indicators of capability which would not be the case in the absence of such processes.

2. Western democracies are uniquely unsuited to safe participation in arms control negotiations.

The virtues of Western-style democracy, which need not detain us in this context, tend to become vices with reference to arms control negotiations with the USSR. Our distinctive strategic culture, which reflects our self-assessment, our geopolitics, and our reading of our history, tends, in usual ethnocentric fashion, to be projected upon a country which, in reality, has an alien strategic culture.[32] The United States and NATO-Europe seek enhanced 'stability', as the central guiding concept through SALT and MBFR, yet the USSR does not acknowledge the legitimacy even of a rough facsimile of the Western ideas that coalesce to form that concept.

The Soviet Union is in the business of becoming still more powerful, possibly in pursuit of the somewhat unimaginative goal of becoming more powerful yet, and her idea of stability is the preservation of a theatre campaign-winning capability in Europe and the improvement of her relative prowess *vis-à-vis* US central war-waging assets. Largely for cultural reasons, Western democracies have, until quite recently, simply ignored the plain, inconvenient, yet obvious facts concerning Soviet strategic culture. The 'satisfied' character of Western democracies explains, in large part, their near obsession with the concept of stability and order; while the very openness of their mass participatory political systems facilitates the interdiction of policy-making processes by charitable explanations of Soviet motives and behaviour.

However, it is only fair to observe that in the United States and Great Britain, at least, wishful thinking concerning Soviet intentions probably is an elite rather than a mass public problem. The potential for bellicosity and insular truculence of the US electorate is a well-known fact which gives pause to all shades of defence opinion. Although 'peace' is always popular (witness Richard Nixon's campaign in 1972), candidates for the Presidency who choose to run against the Soviet Union have a very large, potentially credulous constituency on that subject. In the United States, at least, there is good reason to believe that although detente – interpreted as meaning peace – was popular, there has remained a very fundamental distrust of the Soviet Union which could be tapped fairly easily by a President, or a candidate for the Presidency.

Mr Nixon's 'silent majority' was no figment of the political professional's imagination. Deep down, the US body politic does not trust the Soviet Union, is uncomfortable doing business with the Soviet Union, and is eminently susceptible to the argument that the Soviet Union has been up to no good of late. To the extent to which detente euphoria was a reality in the early-to-mid-1970s, it infected liberal-minded opinion leaders and (otherwise conservative) businessmen; it did not suffuse the American body politic writ large.[33]

The military, and perhaps hence political, problems facing the West in the 1980s were predictable – *and were predicted*[34] – in the early 1970s. However, respected and authoritative institutions such as the IISS, Chatham House and the Council on Foreign Relations, tended to reflect the fashionable temper, and 'responsible opinion', of the times, and did not often accommodate in their publications the kind of warnings that should have been issued. Although none of these institutions are constitutionally permitted *institutional* positions, centres of strategic expertise have, perhaps, been guilty of excessive pandering to the prevailing intellectual fashion and political climate, rather than serving as vehicles for the accurate presentation of trends which, from a sober strategic standpoint, could only be seen as dangerous and potentially destabilising and which have hardly emerged overnight.[35] Despite the due recording of the steady, across-the-board augmentation of Soviet military might, institutes such as the IISS, among others, have abjured in their responsibility to be more prescient than the norm.[36]

In his Department of Defense Annual Report, Fiscal Year 1981, Harold Brown makes the extraordinary admission that

> Critical turning points in the histories of nations are difficult to recognize at the time. Usually they become clear only in retrospect. Nonetheless, the United States may well be at such a turning point today. We face a decision that we have been deferring for too long; we can defer it no longer. We must decide now whether we intend to remain the strongest nation in the world. The alternative is to let ourselves slip into inferiority . . .[37]

Dr Brown's 'critical turning point' came after being in office for a full three years. He did not acknowledge that he and the Administration in which he held cabinet-level responsibility for defence policy, contributed to the slide in relative US capability. The cancellation of the B-1 bomber, and the delaying of Trident submarines and MX missiles, were events that had strategic consequences. It is arguable that many institutions of strategic analysis, although possibly not a part of the problem, have been less than very useful contributors to the search for a solution.[38]

3. Arms control processes, in practice, tend to encourage the redefinition of strategic problems as arms control problems.

Given the fundamental ambivalence of Western policy-makers over the purposes of arms control and the significance of 'details' – by way of contrast to the Soviet General Staff analysts who know that arms control is subordinate to defence planning[39] – it is hardly surprising that negotiability and political appearances often loom larger in policy assessment than does defence planning rationality. The ease with which 'arms control problems' assume a life of their own is explained, in good part, by the neglect of, or disdain for, strategic *operational* issues.[40] A political system which believes deeply that a nuclear war would mean 'the end of history' is not a political system very likely to attend carefully to military operational details. If, in common, say, with Christoph Bertram, one is mightily impressed with the stability and resilience of the European and central strategic balances,[41] and if one believes that East-West war is deterrable through a mixture of some denial capability and a massive ability to punish an enemy's society, then the details of arms control agreements tend to fade in their inherent significance.

Fundamentally, neither NATO-Europe nor the United States believes that it needs to plan to *defend* successfully against Soviet attack, a fact well-reflected both in the 'planned insufficiency' in the European theatre, and in the near absence of war-survival and recovery programmes in the United States. Arms control, *à la* SALT and MBFR, encourages the persuasive fallacy that the Western Alliance is *really* in the deterrence-only business, and that defence (or denial) capability is important only in so far as it strengthens pre-, or early intra-war deterrence.[42]

As with most good arguments, the allegation that the existence of arms control processes has impaired rational defence planning can be taken too far. There is, in fact, a considerably heated debate being conducted in the United States at present over the degree of truth in this allegation.[43] In addition, there is scarcely less heated a debate over the merit in the reverse allegation, that arms control has, or might, improve the possibility of rationality in defence planning. In support of the contention that arms control has, or is very likely to have, a negative impact on defence planning, one could cite some features of the MX missile programme (including, perhaps, its physical size – and hence warhead count – and the planned scale of the deployment), the scale of the planned ALCM deployment,[44] the scale of the planned Trident deployment, and possibly the range allowed NATO's GLCMs. The technical parameters of the MX ICBM and the scale of the LRCM deployment are not trivial matters of detail. They could bear fundamentally upon the pre-launch and penetration survivability of each system, and of other strategic systems, given the survivability synergism among different elements in the strategic force posture.

Naturally, one should beware of monocausal explanations. It may be an exaggeration to call the MX missile 'the son of SALT'.[45] None the less, as Soviet commentators are fond of saying, it is no accident that the planned payload of the MX (at close to 8,000 lb) is the most that can be accommodated within SALT understandings of what is a 'light' ICBM. Overall, it is scarcely controversial to note that in SALT to date, the Superpowers have chosen the least important of the possible accounting units: missile launchers, rather than missiles themselves or payload. Launcher or platform limitations tend almost naturally to drive a defence community to make each permitted launcher or launch platform as individually capable as possible.[46]

It can be argued that the ABM Treaty of 1972 is an aid to rational defence planning in that officials know for certain that ballistic missiles will not be actively opposed. This may or may not be true, depending upon how seriously one assesses Soviet Treaty 'breakout' and rapid deployment possibilities for the 1980s.[47] Some people in the US BMD community believe that the Soviet Union has made as rapid progress in ABM research and development with the Treaty as could have been made in the absence of the Treaty. Quite aside from the issue of just how predictable is a 'BMD-less future', predictability need not be a virtue if it refers to an undesirable condition. As this author has argued elsewhere, BMD, as a component in a fairly balanced offence-*defence* posture, is essential for strategic stability.[48]

The principal support at present for the thesis that arms control can assist rational defence planning makes reference to the alleged dependence of the MX missile upon the payload fractionation sub-limits of SALT II.[49] The argument holds that with SALT constraints, the upper boundary of the Soviet offence-forces threat can be predicted with high confidence, thereby allowing the United States to proceed with an MX deployment design which it knows will provide adequate pre-launch survivability.[50] As many supporters of SALT II have noticed, this is a terrible argument. The predictability offered by SALT II would last only until December 1985 – a period that will not see any MX deployment.[51] No persuasive theory has yet been advanced which attempts to explain how the Soviet Union may be induced to co-operate in a SALT III to the effect that the single most potent weapon in the American strategic arsenal could not be targeted with profit.[52]

Although this discussion of arms control is highly critical in tone and substance, it should not be imagined that arms control processes comprise the principal villain impairing rational defence planning. The real villain, if one must be isolated, is the unwillingness, or inability, of the United States and NATO to think strategically.[53] Unless one has a reasonably clear idea of the political objectives that the application of force should be designed to attain, it is difficult to decide upon a strategy, just as it is next to impossible to decide what forces are necessary. In short, if one has not thought through the strategic connection between posture and war aims, the entire structure of defence planning is unsound. More to the point, given the mandate of this chapter if one lacks a strategy which indicates the 'worth' of particular elements of strategic capability, there can be no strategically rational basis for arms control policy planning.

It may be objected that this author is confusing a strategy that he does not like (in the structure of the current US SIOP) with the absence of strategy. That is not the case. This author is claiming that: first, the US defence community has not thought through the relationship between war aims and force application[54]; and secondly 'defence planning', so-called, is a conceptual aggregate which conceals the fact of the real separation of streams of activity which should be intimately interconnected (targeting, research, development and procurement, declaratory doctrine (often confused with the concept of 'policy'), and arms control policy). US arms policy has been the victim of a defence community that is neither organised, nor inclined, to think strategically.

4. Arms control processes, license bureaucratic 'actors', who are indifferent to, and generally, ignorant of, defence planning concerns, to interdict strategic policymaking.

The existence of interstate arms control processes has the effect of politicising defence programme decisions (which may or may not be desirable in the Soviet context). This tends to produce a disadvantageously asymmetrical situation for the West. In practice, in the 1970s, it has meant that US defence planning has been unusually influenced by banner carriers for self-restraint in arms policy (located in State, ACDA and the NSC staff),[55] while Soviet SALT and MBFR policy has been generated, in most important respects, by the Soviet General Staff.

The diffusion of national security policy-making responsibility in Washington – particularly acute in the Carter Administration – and the very structure of government, tends to polarise official debate on arms and arms control policy. As Richard Burt has argued, it is almost certainly unfortunate that ACDA was created.[56] Such an agency, with an institutional interest in the promotion of arms control, could hardly help but cast itself, as a general rule, in the role of advocate for restraint. In short, both the Joint Chiefs of Staff and ACDA are really driven by bureaucratic circumstance into polarised positions. Also, the proliferation of arms control offices in the Department of State and the Office of the Secretary of Defense (*inter alia*) has tended to drive ACDA and the JCS to adopt extreme, or perhaps pure, policy advocating positions.

The problem is that because of the character of the political rivalry with the Soviet Union, the only measures of arms control which are negotiable tend to be those which are irrelevant (i.e. both sides agreed not to do things which they do not want to do anyway), trivial (i.e. 'constraints' are established which really register and license what both sides want to do over the period in question (SALT I and II)), or harmful (i.e. agreements are signed which have real substance, but which function to the Western disadvantage (the ABM Treaty)). If arms control processes could usefully reduce the risks of war or the likely damage in war, then a strong case could be made for permitting arms control perspectives to interdict defence planning. However, the promise of arms control is so modest that the claims of its bureaucratic advocates for a significant role in national security policy-making are not strong.

The US Defense Department, in and of itself, is ill-equipped to think strategically. That condition is worsened when there are major players in the arms control arena (i.e. the defence planning arena), located beyond its perimeter. Arms control policymaking and defence planning should be co-operative activities, but the master should be defence planning. The prospects of war or peace are most unlikely to be affected significantly by any of the details likely to be negotiated through SALT or MBFR. Yet the quality of defence planning could be critically important for deterrence, or for operational success should armed conflict occur.

The author does not mean to imply that government reorganisation necessarily would lead to an improvement in the quality of public policy. Here it is argued only that arms control processes, and the way in which they are managed by the US Government, have contributed to the decline in relative Western military strength. However, arms control has so contributed because it has been approached, in the West, from the perspective of fallacious theories of arms race and crisis stability.

5. Arms control negotiators are as likely to constrain the wrong force elements as the right ones.

Stated succinctly, the human power of technological, tactical and strategic prediction is not impressive. In a period of rapid technological change, an arms control regime is as likely to do harm to the (Western) goal of stability as it is to do good. Even in the realm of strategic nuclear weaponry, which is vastly less complex for analysis than is the theatre balance in Europe, arms control in the 1970s committed some signal errors. The preclusion of effective hard-point defences through the ABM treaty was a serious error while the US insistence, in 1972, that land mobility for ICBMs would be judged to be incompatible with the SALT I regime, was scarcely less short-sighted. The late Bernard Brodie's observations on the undesirability of arms control in an era of rapid military-technological 'improvement' – in the context of the complex naval competition of the last decades of the nineteenth century and the first decade of the twentieth – merit sympathetic attention.[57]

It is easy to forget that prior to 1972 the East-West arms competition was 'unconstrained' in any formal sense worthy of note. Stability, or equilibrium, may be negotiable through arms control, but it is enforceable only through competition. More pointedly, to risk stating the obvious, yet again, if stability is *not* enforceable through competition – because, for example, the United States, and/or NATO, is unwilling to compete – it is *not* negotiable through arms control.

As this author has argued in detail elsewhere, it is difficult to improve Western security through arms control if one adheres to incorrect or to eminently challengeable strategic theories.[58] The mainstream of thinking within the US defence community has long believed that 'hostages [civilian assets] must remain unambiguously vulnerable and retaliatory forces must remain unambiguously invulnerable'.[59] This mutual vulnerability axiom has served as the basis for theories of arms-race and crisis stability. The 1970s have shown that the dynamics of the strategic arms competition cannot be explained on the basis of this theory. The Soviet Union does not accept the idea that it is desirable for both Superpowers to be unambiguously vulnerable to retaliation. Moreover, there is something to be said for the argument that a measure of strategic instability may be beneficial for political stability (which is all important *vis-à-vis* decisions to fight or not to fight).

Overall, the case for arms control, as a potentially stabilising factor, remains to be proven. The alternative to arms control is not a 'runaway' arms race; there will always be resource constraints and constraints stemming from calculations of anticipated net (dis)advantage. If the urge to engage in some facsimile of a crash rearmament programme were very strong indeed, then no extant arms control regime could survive.

6. The separation of so-called strategic arms in SALT from other kinds of defence capability has undermined the very basis of NATO strategy.

In extremis, NATO security rests on the willingness of a US President to lead a process of nuclear escalation in, and beyond, the European theatre. Stability (through mutual vulnerability) at the strategic nuclear level, a goal of arms controllers in the 1970s, probably is incompatible with the premisses of Western European security.[60] If both Superpowers have secure counterdeterrents and unprotected homelands, how can the initiative be taken? SALT encouraged the evolution of this strategic problem because it was accepted widely in the United States that acknowledgement of 'the parity principle' would have to be the basis of the deal to be struck. Strategic rough parity or 'essential equivalence' could be accommodated by NATO, provided the theatre balance were corrected. However, the complex theatre balance – conventional, nuclear and relevant naval – deteriorated through the 1970s.

The current crisis of arms control flows, in minor part, from the following apparent fact: conveniently manageable slices of the total problem of East-West military competition cannot be isolated for treatment, because each slice lacks integrity considered in isolation. This is a descriptive point – the United States and NATO could change their posture and doctrine so as to facilitate a piecemeal approach to arms control problems. However, pending correction of the theatre balance, 'essential equivalence' is a prescription for strategic paralysis. According to Harold Brown, 'it enhances stability in a crisis by reducing the incentives for either side to strike first or preempt'.[61] This is a vision of a 'technological peace' apparently untroubled by the fact that NATO's hopes for a continuity in deterrent effect rest substantially upon the credibility of first US use of central systems.

Unfortunately for the future of arms control, it is not obvious that the evident deficiencies in the mandates of SALT and MBFR – with regard to their fit with NATO strategy – could be corrected through some grand aggregation of arms control concerns in a new arms control forum with a far more extensive mandate. The policy rationality of a single arms control forum would be more than offset by the complexity of the negotiating items and the diversity of national interests that would have to be represented.

Although one can sustain the argument that the diplomatic structure of East-West arms control in the 1970s – with reference to the MBFR 'guidelines area', and the separate treatment of so-called strategic forces – was contrary to Western security interests, there can be no evading the fact that the real problem was a deficiency in Western military programme momentum.[62] No matter how cleverly one seeks to reorganise the diplomatic institutions of arms control, there can be no substitute for trained men and weapons (and appropriate operational doctrines). The Soviet Union cannot be persuaded to forgo the benefits of military preponderance in the European theatre, in the interest of a Western theory of stability.

7. Western arms control negotiators have been like surgeons operating in the absence of a prior diagnosis of the disease they are treating.

It is tempting to argue that the arms control processes of the 1970s failed in good part because of Western misunderstanding of the nature of the arms control/arms race problem. However, that temptation should be resisted because it is more likely than not that arms control would have failed, or more likely would not have been attempted, had the true nature of the problem been comprehended. The complex East-West arms competition is a competition between satisfied Powers and a revolutionary Power. This apparently unsophisticated, even naïve, observation has profound policy implications: indeed, so profound that they escaped official attention until quite recently. Contrary to the sense of the confident shibboleths of ten years ago, the East-West arms competition does not appear to be fuelled by a mechanistic action-reaction process, wherein each side responds, often in anticipation, to the military threat perceived abroad to the assured destruction capability identified as *the* touchstone of an adequate deterrent. Instead it is now recognised fairly widely that there is a very large measure of autonomy in both the scale and detail of each side's competitive activity,[63] and that the Soviet Union appears to be moved by an enduring determination to achieve a war-waging/war-winning capability.[64]

In order to control the arms competition, one could think that it might be useful to have a good idea as to how the arms race system 'worked'. In the 1970s Western negotiators assumed a reactive arms race dynamic that simply did not exist. This is hardly surprising, for as Michael Howard has suggested: 'Works about nuclear war and deterrence normally treat their topic as an activity taking place almost entirely in the technological dimension. From their writing not only the sociopolitical but the operational elements have quite disappeared.'[65]

As best can be determined from the evidence available, Soviet arms competitive activity can be explained in very large part by reference to 'sociopolitical' and 'operational' considerations, and scarcely at all by reference to 'the technological dimension' that pervades theoretical American arms control writings. Stated somewhat crudely, the USSR performs as she does in military matters very largely because of the kind of country that she is. The theory of arms-race stability (via mutual societal vulnerability), as deployed over the past ten years to condemn urban-industrial BMD, MIRV, the Mk 12A Re-entry Vehicle, and MX, has been strong on *the* (really *a*) logic of technology, but heroically weak on strategic-cultural empathy.

In Soviet perspective, a commitment to defend the homeland, not to place any voluntary reliance on the restraint of others, and to seek military preponderance, are really beyond the realm of negotiability. This is not to say that useful arms control business cannot be conducted with the USSR. Rather it is to claim that Western arms control policies should begin with recognition both of the inalienable requirements of Western security (appraised in an *operational*, rather than pre-war deterrent, perspective), and of the enduring features of Soviet strategic culture.

It is no exaggeration to claim that the mainstream of American arms control advocates approached the SALT negotiations with little, if any, appreciation of the fact that their vision of strategic stability – which they thought could be registered, or at least, advanced, through SALT I[66] – had potentially revolutionary implications for the stability of the Soviet state. The CPSU is the vanguard of a proletariat which seized power in the weakest link of capitalism. The idea of a 'technological peace', reflecting the mutual deterrence impact of nuclear weapons, must be fundamentally challenging to the legitimacy of the Soviet state. The Soviet Union cannot genuinely acknowledge the vital interests of antagonistic social systems. The Soviet-American, and more general East-West, multi-level arms competition is both fuelled by, and is about, *politics*. The arms control technicians of Western defence communities have persisted in believing that the arms competition is fuelled by military-technical triggers.

It is tempting to believe that although arms control, to date, shows a very meagre, or even negative, record of accomplishment, more careful research, more imaginative thinking, or simply more competent diplomatic management, will produce much better results. This may be correct, but there are reasons for scepticism. The movement for the reform of arms control is truly ecumenical, embracing left-, centre-, and right-oriented analyses. These analyses tend to begin with Leslie Gelb's judgement, quoted at the beginning of this chapter, that 'arms control essentially has failed'. But what remedies typically are proposed?

Change the focus of constraint from input *(launch platforms, warheads, etc.) to strategic* output. In a powerful analysis, Christoph Bertram of the IISS has argued for the constraint of (undesirable) strategic 'missions'.[67] This is unexceptionable, save for the considerations that there is an absence of consensus over which missions are 'destablising' and which are not; and that the 'missions' which Dr Bertram is most concerned to constrain are precisely those dearest to the heart of the Soviet military-political establishment. So, in order to be taken very seriously, such a proposal first needs to inform us as to how to effect a radical change in Soviet strategic culture.

Seek to achieve much less. This view can come close to the proposition that an arms control process may have considerable value simply by virtue of its very existence. From time to time, this process may bring forth very modest agreements, but these would not have the status of being major diplomatic events. This apparently sensible, centrist option fits well the evidence of the past decade; strong sceptics are advised to read, or re-read, Richard Burt's 1978 article which carried the subtitle of 'The Risks Of Asking SALT To Do Too Much.'[68] However, no matter how true it may be to claim that only very modest gains can be registered through arms control, it remains to be demonstrated that the US body politic could live contentedly with such a proposition. As noted already, in this chapter, the very fact of an ongoing arms control process tends to have distorting effects upon the course of defence planning, while the domestic politics of debate over arms control agreements do not lend themselves easily to modest statement. In short, the kind of modesty in supportive rhetoric suggested by this option, is the kind of modesty likely to persuade many potential senatorial supporters that the agreement in question is of so little value that there would be next to no political cost in turning it down.

Compete much more vigorously so as to provide US negotiators with the bargaining chips necessary for the securing of genuinely balanced agreements. This thesis has much to recommend it, but it has to be admitted that it may be so discouraging to Soviet defence planners that they would decline to play.[69] In other words, this is a prescription for the restoration of strategic stability (on terms acceptable to Western defence communities), but it is not necessarily a prescription for the rejuvenation of the arms control process.

Strong-minded and broad-fronted assaults on arms control agreements and processes lend themselves to easy misinterpretation. To return to basics, the fundamental goals of arms control outlined earlier are not in question. This author does not believe that the arms control processes of the 1970s contributed usefully to the achievement of any of these goals.

Arms Control in Europe?

Although Christopher Makins is correct in his assertion that the US, British and West German governments did not take MBFR very seriously in the early 1970s (or since),[70] it would be an error to believe that a truly serious approach would likely have produced more substantial an outcome. Disagreeable though it may be to recognise, the plain fact is that European arms control – compatible, that is, with NATO-European security interests – probably has no plausible future. MBFR has not 'failed', thus far, because of inattention by policy-makers, lack of imagination, or because of its umbilical tie to other (vanished or rapidly departing) detente processes. MBFR has 'failed', if that is the correct expression, because there is an insufficient community of political interests linking East and West in the enterprise.

The Soviet Union left the 1970s, as she entered them, militarily superior in Europe.[71] The only negotiable option offered NATO by the Soviet Union in MBFR was the ratification, at best, of that superiority. More seriously, as Lothar Ruehl has indicated,[72] NATO might just have signed an MBFR regime which would have constrained its defensive potential very markedly. There is nothing surprising in this. The Soviet Union, reasonably enough, is content with the military (im)balance in Europe as it has evolved over the past decade. NATO, to date, has provided the Soviet Union with no persuasive incentive to negotiate a militarily more balanced stand-off in Central Europe. *The First Law of Arms Control states that you can only negotiate about that which you have demonstrated an ability and willingness to achieve unilaterally*.

It is difficult to know whether to laugh or cry when some NATO-Europeans insist that NATO's long-range theatre-nuclear force modernisation programme must be linked to a NATO-Warsaw Pact arms control process.[73] Europeans, schooled by history in *realpolitik*, are supposed to be wise in the ways of the world. How can it be 'provocative' for Dutch-deployed GLCMs (for example) to be targeted on the USSR, when Soviet SS-20s are already targeted on the Netherlands? To be blunt: in the name of detente, peace and arms control, the reality of naïveté, cowardice and even treason[74] is taken seriously, as a domestic political fact, in NATO councils.

This is not to defend NATO's LRTNF modernisation programme uncritically. It is not obvious that the most effective way to place the Soviet second echelon at risk is through the threat of Pershing II and GLCM strikes. Moreover, if NATO is seeking to respond *directly* to the SS-20 threat, the planned LFTNF modernisation programme is irrelevant. Since the SS-20 is a truly mobile system, effective responses would be deep-penetrating strike aircraft (to *find* the SS-20) and/or ballistic missile defence. Given its relatively modest throw-weight and re-entry velocity, the SS-20 lends itself to BMD interdiction. BMD, of course, evades the NATO-European political difficulty attending weapon systems that could strike at Soviet soil. This suggestion is offered in full recognition of the problems pertaining to the details of the ABM treaty, and British and French anxieties concerning the ease of penetration of their ballistic missiles.[75] Britain and France would be right to worry about the implications for the penetration survivability of their ballistic missiles of a Soviet-American BMD competition. However, there would be much to be said in favour of a NATO-Europe virtually immune to SS-20 strikes, and a US strategic force that could not be targeted intelligently.

Arms control in Europe has been a chimera from the very beginning (which, given the mandate of this chapter, may be dated at October 1957 with the so-called 'Rapacki Plan' for disengagement).[76] NATO-European, and particularly continental NATO-European states have been guilty, in recent years, of misleading their electorates on the subject of arms control and disarmament in Europe. The reason for this rests with geopolitics: it is a geographical fact that the Soviet Union is a European Power, while the United States can only be a Power in Europe. There could never be a European arms control agreement which truly would register (a Western-style) stability in Central Europe, given the fact of the permanent Soviet presence in the theatre. Moreover, the Soviet Superpower presence in Europe effectively has discouraged NATO-Europe from seeking to establish even a tolerable facsimile of an intra-theatre military equilibrium. As noted earlier, the 'planned insufficiency' in theatre forces *should* guarantee a robust linkage between the US strategic nuclear posture and NATO's Central Front.

So long as NATO seeks to insist, through the MBFR forum, that the disadvantageous *status quo* be changed dramatically in its favour by arms control agreements, then so long must MBFR remain moribund. Any NATO-European politician who refers optimistically to the prospect for MBFR, or other European-theatre arms control, agreement has to be either intellectually foolish or engaged in political posturing. The existing, and evolving, European military (im)balance is very congenial to the USSR. For excellent Soviet reasons, there is no prospect whatsoever that an MBFR agreement with genuinely stabilising features might be negotiable – barring, that is, an extended period of NATO defence refurbishment.

Conclusion

It is worth mentioning that the arms control processes of the 1970s might well have no future in the 1980s. At the very least it is unlikely that SALT will rise again for several years. Arms control business cannot be conducted in an atmosphere of acute political suspicion. With the SALT process at least temporarily defunct in the United States, there can be no movement in MBFR. In the half-light cast by the evidence of Soviet misbehaviour in Afghanistan, perhaps NATO-Europe will recognise the postulated linkage between LRTNF modernisation and East-West arms control for the military-political folly that it always was. However, the grounds for optimism on that score are less than modest.

In Europe, or elsewhere, the Soviet idea of stability translates into the concept of preponderance. NATO's MBFR planning needs, finally, to be shed of the aspiration that the USSR might be persuaded, or negotiated, out of local advantages that NATO judges to be destabilising. A stable situation in Europe, in Soviet perspective, is one where they can seize and retain the military initiative. NATO need not, and should not, accept that definition of stability.

Notes

1. A point well made and illustrated in Christopher J. Makins, 'Negotiating

European Security: The Next Steps', *Survival*, vol. XXI, no. 6 (November-December 1979), pp. 256-63.

2. Christoph Bertram, *The Future of Arms Control: Part II: Arms Control and Technological Change: Elements of a New Approach*, Adelphi Papers no. 146 (IISS, London, Summer 1978), p. 30.

3. Leslie H. Gelb, 'A Glass Half Full', *Foreign Policy*, no. 36 (Fall 1979), p. 21.

4. Former Secretary of State Cyrus Vance has implicitly criticised President Carter strongly for his decision to defer proceeding with a Senate floor debate on the subject of SALT II ratification. 'Text of Vance Speech at Harvard on Foreign Policy', *The New York Times*, 6 June 1980, p. A12. Mr. Vance was moved to observe that 'it is far too easy, in an election year, to let what may seem smart politics produce bad policies'.

5. It is too early to know whether this hiatus will bring forth an interesting theoretical literature. However, the judgements in this paragraph reflect the author's experiences at meetings in the United States and Europe in the first half of 1980.

6. For 'classical' statements of what arms control might accomplish, see Donald G. Brennan, 'Setting and Goals of Arms Control', *Daedalus*, vol. 89, no. 4 (Fall 1980); and Hedley Bull, *The Control of the Arms Race: Disarmament and Arms Control in the Missile Age* (Weidenfeld and Nicolson, London, 1961), Chapter 1. For a mid-1970s reassessment, see Bernard Brodie, 'On the Objectives of Arms Control', *International Security*, vol. 1, no. 1 (Summer 1976), pp. 17-36.

7. For examples see US Senate, Committee on Foreign Relations, *The SALT II Treaty, Hearings*, Part 1, 96th Cong., 1st Sess. (US GPO, Washington, 1979), pp. 554-5, 588-9.

8. The Soviet invasion of Afghanistan almost certainly reflected the extremely limited interest the USSR retained in the consummation of SALT II. As this author, and others, have argued, SALT II is not substantively very important. Clearly, the Soviet Union agrees. See the articles by Gregory F. Treverton and Colin S. Gray in *Survival*, vol. XII, no. 5 (September-October 1979), pp. 194-7 and 202-5, respectively.

9. Probably the most powerful critique, overall, is Robin J. Ranger, *Arms and Politics, 1958-1978; Arms Control in a Changing Political Context* (Macmillan, Toronto, 1979).

10. See Edward L. Rowny, 'Negotiating with the Soviets', *The Washington Quarterly*, vol. 3, no. 1 (Winter 1980), pp. 58-66.

11. For example, see the letter from Paul H. Nitze to Senator Church reproduced in US Senate, Committee on Foreign Relations, *SALT II Treaty, Hearings*, Part 1, pp. 529-30.

12. On the questions of what is 'stability', and how it might be encouraged, see my essay 'Strategic Stability Reconsidered', *Daedalus* (Fall 1980).

13. NATO-European security rests very largely upon the health in the trans-atlantic political connection. In European eyes, deterrence rests far more upon a Soviet belief that the United States is vitally concerned with developments in Western Europe, than upon estimates of relative putative East-West military prowess.

14. Note the judgement to this effect in Peter Nailor, 'The Strategic Context', in Peter Nailor and Jonathan Alford, *The Future of Britain's Deterrent Force*, Adelphi Papers No. 156 (IISS, London, Spring 1980), p. 17.

15. See Johan J. Holst, 'Strategic Arms Control and Stability: A Retrospective Look' in Holst and William Schneider, Jr. (eds.), *Why ABM? Policy Issues in the Missile Defense Controversy* (Pergamon Press, New York, 1969), Chapter 12; and Ranger, *Arms and Politics, 1958-1978, passim*.

16. Brennan, 'Setting and Goals of Arms Control', pp. 40-1.

17. 'Central strategic war, according to Soviet literature is not likely to stem from mechanistic instabilities within the super-power military relationship, but rather from real and enduring differences between competing political systems and national interests.' Richard Burt, 'Arms Control and Soviet Strategic Forces: The Risks of Asking SALT to do Too Much', *The Washington Review*, vol. 1, no. 1 (January 1978), p. 22.

18. This is the title of Chapter 9 in Thomas C. Schelling, *The Strategy of Conflict* (Harvard University Press, Cambridge, Mass., 1960).

19. A very useful history of the SALT II negotiations is Strobe Talbott, *Endgame: The Inside Story of SALT II* (Harper and Row, New York, 1979). See also Thomas W. Wolfe, *The SALT Experience* (Ballinger, Cambridge, 1979), particularly Chapters 10-11; and John Newhouse, 'Reflections: The SALT Debate', *The New Yorker*, 17 December 1979, pp. 130-66.

20. West Germany, feeling more exposed than other countries to the chill winds that could blow as a consequence of deteriorating Soviet-American relations, has played divisible detente politics with East Germany and even with the USSR (notwithstanding Bonn's decision not to send a West German team to the Moscow Olympics) of recent months, despite the fact that her vulnerability renders her exceptionally dependent upon US security guarantees.

21. Recent unilateral West German *démarches* with respect to theatre-nuclear arms control proposals, have come close to or even exceeded, the margin of US tolerance.

22. The plain fact is that the USSR sees stability in her extant measures of preponderance within the European theatre. Soviet strategic culture sees no merit in the idea of balance of equivalence – it has an unambiguous 'battlefield' orientation.

23. See Henry Kissinger, 'The Future of NATO', *The Washington Quarterly*, vol. 2, no. 4 (Autumn 1979), pp. 3-12.

24. John Newhouse claimed that 'the talks [SALT] were launched not from a common impulse to reduce armaments, *but from a mutual need to solemnize the parity principle* – or, put differently, to establish an acceptance by each side of the other's ability to inflict unacceptable retribution in response to a nuclear attack'. (Emphasis added). See *Cold Dawn: The Story of SALT* (Holt, Rinehart and Winston, New York, 1973), p. 2.

John Erickson, 'The Soviet View of Nuclear War', talk broadcast on BBC Radio Three, 26 June 1980. This operational approach may be observed in action in the details of Soviet defence preparation, and is advertised very openly in the serious Soviet military literature. For example, between 1965 and 1975, the Military Publishing House of the Ministry of Defence of the USSR produced a set of seventeen (text) books which comprise the 'Officer's Library'. The purpose of these books is 'to arm the reader with a knowledge of the fundamental changes which have taken place in recent years in military affairs'. Col. Gen. N.A. Lomov, (ed.), *Scientific-Technical Progress and the Revolution in Military Affairs (A Soviet View)*, Soviet Military Thought Series, no. 3, trans. by the US Air Force (US GPO, Washington, 1975) (first publ. Moscow, 1973), p.V. The more important books in the 'Officer's Library', as with the Lomov volume, are available in the US Air Force's 'Soviet Military Thought' series. A careful Western study, resting exclusively on Soviet sources, is Joseph D. Douglass, Jr., and Amoretta M. Hoeber, *Soviet Strategy for Nuclear War* (Hoover Institution Press, Stanford, 1979).

26. I have addressed this question in my article, 'Arms Control in Soviet Policy', *Air Force Magazine*, vol. 63, no. 3 (March 1980), pp. 66-71.

27. On the concept of strategic culture, see Jack L. Snyder, *The Soviet Strategic Culture: Implications for Limited Nuclear Operations*, R-2154-AF (RAND, Santa Monica, September 1977).

28. The 'causes of war' literature is vast and, scarcely surprisingly, heroically inconclusive. Useful treatments include Geoffrey Blainey, *The Causes of War* (The Free Press, New York, 1973); and Keith L. Nelson and Spencer C. Olin, Jr., *Why War? Ideology, Theory, and History* (University of California Press, Berkeley, 1979).

29. See Colin S. Gray, 'The Urge to Compete: Rationales for Arms Racing', *World Politics*, vol. XXVI, no. 2 (January 1974), pp. 207-33.

30. An arms control context tends to encourage the simple-minded view that individual weapons may be classified as being either 'stabilising' or 'destabilising'. See Jerome H. Kahan, *Security in the Nuclear Age: Developing US Strategic Arms Policy* (Brookings, Washington, D.C., 1975), pp. 272-3.

31. A strong case can be made for returning to the early mid-1960s idea of pursuing arms control goals through informal processes. In this connection, considerable value may be found in Jeremy J. Stone, *Strategic Persuasions: Arms Limitations Through Dialogue* (Columbia University Press, New York, 1967).

32. See Ken Booth, *Strategy and Ethnocentrism* (Croom Helm, London, 1979).

33. Although it is true to claim that the American public expects the President to offer guidance on the subject of foreign peril, it is no less true to claim that excessive expectations of detente characterised the foreign policy opinion leaders in the West, not the general public (opinion polls in the United States through most of the 1970s show a consistent pattern of deep-rooted popular distrust of the USSR).

34. Such authorities as Adam Ulam, Richard Pipes and William Van Cleave were making the same pessimistic assessments concerning Soviet intentions in 1970-72 that they are making today. Their judgements were correct – on the evidence of the 1970s. Somewhat later, the group of conservative Soviet experts who performed an adversary assessment role as 'Team B' in 1976, second-guessing the official (US) national intelligence estimating community, again, were proved by events to have been very substantially correct in their prognoses.

35. On the putative Soviet 'window of opportunity' in the 1980s, see 'Kissinger's Critique', *The Economist*, 13 February 1979, pp. 17-22; and Edward N. Luttwak, 'After Afghanistan, What?', *Commentary*, vol. 69, no. 4 (April 1980), pp. 40-9.

36. For example, as late as the summer of 1979, the IISS chose as its annual conference theme the subject of 'The Future Strategic Deterrence'. The speeches and papers at that conference essentially reflected a fairly relaxed, mid-1970s view of the world.

37. *Department of Defense Annual Report, Fiscal Year 1981* (US GPO, Washington, 29 January 1980), p. 14.

38. It is easy to be misunderstood. This author is not convinced either that the IISS, for example, could, consistent with its charter, have done as I suggest, or that Western policy necessarily would have been altered had the IISS managed to issue judicious warning. My point is rather to the effect that the IISS did not appear even to try to focus public attention upon what became, as the 1970s proceeded, a more and more plausibly dangerous prospect for the 1980s. In other words, this is to criticise the IISS not for sins of commission, but possibly for sins of omission.

39. See Wolfe, *The SALT Experience*, Chapter 3.

40. A point most tellingly made in Michael Howard, 'The Forgotten Dimensions of Strategy', *Foreign Affairs*, vol. 57, no. 5 (Summer 1979), particularly pp. 981-6.

41. Bertram, *The Future of Arms Control: Part II: Arms Control and Technological Change: Elements of a New Approach*, pp. 6-7.

42. I have explored the fallacies in this orthodox Western approach in 'Targeting Problems for Central War', *Naval War College Review*, vol. XXXIII, no. 1 (January-February 1980), pp. 3-21; and (with Keith Payne) 'Victory is Possible', *Foreign Policy*, no. 39 (Summer 1980), pp. 14-27.

43. Richard Burt of *The New York Times* was the most vociferous proponent of this charge. Burt has advanced the argument, in detail, in several conference papers, including that reproduced in this volume.

44. On 120 B-52s.

45. A description favoured by my late colleague, Donald G. Brennan.

46. It is worth recalling that in their classic study, *Strategy and Arms Control* (The Twentieth Century Fund, New York, 1961), p. 2, Thomas Schelling and Morton Halperin observed that 'it [arms control] may by some criteria seem to involve more armaments not less'. The launcher limits of SALT almost certainly have been a serious mistake – viewed from the perspective of arms control objectives. In a MIRV era one cannot sensibly seek crisis stability through constraints on MIRV launcher numbers.

47. See Carnes Lord, 'The ABM Question', *Commentary*, vol. 69, no. 5 (May 1980), pp. 31-8; and Daniel Goure and Gordon H. McCormick, 'Soviet Strategic Defense: The Neglected Dimension of the US – Soviet Balance', *Orbis*, vol. 24, no. 1 (Spring 1980), pp. 103-27.

48. Colin Gray, 'A New Debate on Ballistic Missile Defence', *Survival*, March/April, 1981.

49. These are contained in the First Agreed Statement pertaining to paragraph 10 of Article IV of the SALT II Treaty. See *SALT II Agreement, Vienna, June 18, 1979, Selected Documents* no. 12A, Bureau of Public Affairs, US Department of State (US GPO, Washington, D.C., June 1979).

50. See Leslie H. Gelb and Richard H. Ullman, 'Keeping Cool at the Khyber Pass', *Foreign Policy*, no. 38 (Spring 1980), p. 7.

51. The planned initial operational capability (IOC) date for MX is July 1986.

52. There is a great deal of merit in the argument in William H. Kincade, 'Will MX Backfire', *Foreign Policy*, no. 37 (Winter 1979-80), pp. 43-58. Although Kincade's belief that MX will prove to have a destabilising effect on the arms competition is, I believe, false, he is certainly correct in pointing to strong elements of wishful thinking on the part of some official defenders of the programme.

53. This case is argued, for example, in Edward N. Luttwak, 'SALT and the Meaning of Strategy', *The Washington Review*, vol. 1, no. 2 (April 1978), pp. 16-28; 'The American Style of Warfare and the Military Balance', *Survival*, vol. XXI, no. 2 (March-April 1979), pp. 57-60; and in Colin S. Gray, *Strategy and the MX* (The Heritage Foundation, Washington, D.C., June 1980).

54. See my article, 'Nuclear Strategy: The Case for a Theory of Victory', *International Security*, vol. 4, no. 1 (Summer 1979), pp. 54-87.

55. See Richard Burt, 'A Glass Half Empty', *Foreign Policy*, no. 36 (Fall 1979), pp. 40-2.

56. Ibid., and see note 43. For his most recent sally on this subject, see Richard Burt, 'Its Arms Pact on Shelf, US Agency [ACDA] suffers Ennui', *The New York Times*, 11 June 1980, p.A2.

57. Bernard Brodie, *Sea Power in the Machine Age* (Princeton University Press, Princeton, New Jersey, 1941), pp. 252-6.

58. Gray, 'Strategic Stability Reconsidered', *passim*.

59. Ian Smart, *Advanced Strategic Missiles: A Short Guide*, Adelphi Papers no. 63 (IISS: London, December 1969), p. 28.

60. A point suggested strongly in Kissinger, 'The Future of NATO'.

61. See Harold Brown, *Department of Defense Annual Report, Fiscal Year 1979* (US GPO, Washington, 2 February 1978), p. 57.

62. A point well made in Richard Burt, 'The Scope and Limits of SALT', *Foreign Affairs*, vol. 56, no. 4 (July 1978), p. 761.

63. For a useful review of recent arms race research, see Kendall D. Moll and Gregory M. Luebbert, 'Arms Race and Military Expenditure Models: A Review', *The Journal of Conflict Resolution*, vol. 24, no. 1 (March 1980), pp. 153-85.

64. See John Erickson, 'The Soviet Military System: Doctrine, Technology and "Style" ', in Erickson and E.J. Feuchtwanger (eds.), *Soviet Military Power and Performance* (Archon, Hamden, Conn., 1979), Chapter 2; and Joseph D. Douglass, Jr., *Soviet Military Strategy in Europe* (Pergamon, New York, 1980).

65. Howard, 'The Forgotten Dimensions of Strategy', p. 982.

66. This thesis pervades Newhouse, *Cold Dawn*.

67. Bertram, *The Future of Arms Control: Part II: Arms Control and Technological Change: Elements of a New Approach*, particularly pp. 17-31.

68. See note 17. Also of importance are: Richard Burt, 'Reducing Strategic Arms at SALT: How Difficult, How Important?' in Christoph Bertram (ed.), *The Future of Arms Control: Part 1: Beyond SALT II*, Adelphi Papers no. 141 (IISS, London, Spring 1978), pp. 4-14; and Gelb, 'A Glass Half Full'.

69. William Kincade argues this case effectively in 'Will MX Bacfkire?'

70. Makins, 'Negotiating European Security: The Next Steps', p. 256.

71. For strategic reasons well explained in Douglass, *Soviet Military Strategy in Europe*.

72. In 'The Slippery Road of MBFR', *Strategic Review*, vol. VIII, no. 1 (Winter 1980), pp. 24-35.

73. See 'NATO and Nuclear Force Modernization', *Backgrounder*, no. 110 (The Heritage Foundation, Washington, 14 February 1980), for an excellent summary of the politics of this linkage.

74. This is a strong word – though carefully chosen. It is a matter of public record (reluctantly admitted), to take but one case, that the Dutch Communist Party (CPN), in its lobbying endeavours against the NATO modernisation plan, enjoyed considerable support supplied by the World Peace Council (a well-known Soviet International 'front' organisation).

75. See the discussion in Nailor and Alford, *The Future of Britain's Deterrence Force*, pp. 22-3.

76. See Michard Howard, *Disengagement in Europe* (Penguin, Harmondsworth, 1958).

6 THE FUTURE OF STRATEGIC BALANCE

Desmond J. Ball

In recent years there has been increasing concern regarding the future stability of the global strategic balance. The counterforce capabilities of both the United States and the Soviet Union are improving significantly, threatening the survivability of fixed strategic installations and challenging the role of the land-based components of national deterrent forces. Concern has been voiced in the United States that various developments in the Soviet Union are eroding the American capability to inflict 'unacceptable damage' on the Soviets in any retaliatory strike, and the possibility has been raised of the development and deployment of some quite exotic strategic defensive technologies capable of negating the 'assured destruction' principle upon which mutual deterrence is based. Not for many years has the technological situation been as volatile as it is today, with many novel technologies under active development and many others not much further down the 'pipeline'.

Since it takes from five to seven years for a weapons system concept to be translated into deployable hardware, and a similar period for it to be deployed in sufficient numbers to have any significant impact on the strategic balance, the systems which cause concern range from production-line items (e.g. the large throw-weight Soviet SS-18 ICBM and highly accurate American MIRVs), through prototypes (e.g. the American cruise missiles and the Soviet hunter-killer satellites), to gleams-in-the-eyes of the research and development community (e.g. particle beam weapons).

These technologies, or at least the more proximate of them, do form something of a pattern: the technical future, at least most immediately, clearly lies with greater accuracy, with damage-limitation, and with counterforce capabilities (or at least potential). This is of course, not completely accidental. In recent years both the United States and the Soviet Union have consciously pursued a number of advanced weapons technology programmes and associated support system projects intended to provide alternative options to a purely 'assured destruction' strategy – technologies which permit counterforce targeting, damage-limiting strategies, and a greater nuclear war-fighting capability.

Whether or not any specific development arises simply from follow-on technological imperatives or as a response to the requirements of policy has an important bearing on the relative success of attempts to control these developments and their impact on stability and deterrence. This chapter does essentially three things: first, it describes the recent technological developments in the strategic area, both the spectacular and the less salient but equally consequential; secondly, it attempts to put these developments in some perspective; and, thirdly, it attempts to predict how these developments are likely to affect prevailing notions of deterrence and strategic stability.

Recent Developments in Strategic Technologies

Although it is spectacular technological developments which dominate the public strategic debate, it is sometimes the case that the most significant implications for the strategic balance arise from developments which are insufficiently discrete to command widespread attention. The A-bomb, the H-bomb, and the intercontinental ballistic missile were obviously revolutionary developments. But the development of, for example, low-yield nuclear weapons, of highly accurate ballistic missile re-entry systems, of 'controlled response' command and control systems, etc., while lacking such saliency, has also had profound implications for deterrence, international strategic stability and arms control.

At the more spectacular end of the spectrum, recent debate has focused on work on directed-energy weapons, both charged-particle beam devices and high energy lasers. Claims have been made to the effect that the Soviet Union is building a large experimental particle beam weapon at Saryshagan in Kazakhstan which, when operationally deployed, would be capable of neutralising the entire US ballistic missile force.[1] It would be imprudent to dismiss these claims out of hand. Both the United States and the Soviet Union are investigating concepts involving the military employment of charged particle beam devices, and are developing high-energy technological components that could be used to produce a charged-particle beam weapon. However, several of the component technologies are many years from development (such as the mechanism for steering the beam through the atmosphere and the earth's magnetic field), making it unlikely that any such beam weapon will be constructed or tested before the 1990s.

On the other hand, it is possible that high-energy lasers will be deployed as actual weapons systems somewhat earlier. In particular, both the United States and the Soviet Union are working on technology for space borne laser anti-satellite systems which, if successful, could be operational by the late 1980s.[2] In April 1980, after a moratorium of two years, the Soviet Union resumed active testing of interceptor satellites. The test of 18 April was the seventeenth since the programme began in 1968; ten of these tests were evidently successful.[3] These tests have included successful intercepts at altitudes of over 100 km., which brings the US Navy's Transit navigation spacecraft, in addition to the lower altitude photographic reconnaissance and electronic intelligence (ELINT) satellites, within range of the Soviet killer satellite weapons. However, the Soviets have not yet demonstrated a capability of intercepting satellites at higher altitudes, including, most particularly, the geostationary altitude of 22,300 miles at which are positioned critical components of the American early warning and communications satellite systems.

The United States has responded to this Soviet effort with a number of programmes designed to improve the survivability of its satellite systems. These include the fitting of warning sensors on the satellites; the hardening of sensitive components to protect them against radiation effects; the reduction of satellite radar cross-sections; the deployment of so-called dark or hidden satellites in extremely high orbits that cannot be detected with existing means if all on-board power is turned off until the system is activated upon command; the acquisition of silo-emplaced replacement satellites capable of being launched rapidly in case of emergency; and the development of on-board defensive weapons including a laser for 'killing' enemy killer satellites before they can destroy the American satellites.[4]

Of more significance to the future of the strategic balance, however, at least in terms of the next 10 to 15 years, are a number of less spectacular developments in strategic technologies – in engines, in warheads and in guidance systems. The greatly enhanced precision guidance capacities now available or in the pipeline offer extraordinary accuracy. These guidance systems rely for homing upon either those characteristics of the target which distinguish it from the surrounding environment (by, for example, its optical, infra-red, or radio-wave, etc., signatures), or highly accurate navigation to impact upon fixed targets with known locations or mobile targets passing known locations (by, for example, terrain contour matching, advanced inertial navigation systems, navigation satellites, etc.). These latter methods are principally relevant to missile applications.

American ballistic missile guidance systems remain a generation ahead of their Soviet counterparts. Estimates of Circular Errors Probable (CEPs) vary somewhat according to the nature of the source –

whether, for example, the manufacturer of the guidance systems, the Strategic Air Command or the US Navy, the defence think-tanks, the Pentagon, or the intelligence community. However, it is believed that the current CEP of the Minuteman III ICBM (with the Improved NS-20 guidance system) is as low as 600 to 700 feet.[5] The Advanced Inertial Reference Sphere (AIRS), a 10.75 inch-diameter gimbal-less inertial guidance system being developed for the US Air Force's MX ICBM, is expected to produce a CEP of around 300 feet, or perhaps even 200 feet, probably the limit attainable with purely inertial systems.[6] Although the United States is investigating the application to ICBMs of various techniques of terminal homing, it is not clear, given these very low CEPs, that they will be worth deploying.

Soviet ballistic missiles have also become markedly more accurate over the last decade. In September 1974, the US Department of Defense stated that it had

> some information that the Soviets have achieved or will soon achieve, accuracies of 500 to 700 metres with their ICBMs. These figures may be a little optimistic, but that would represent about a fourth to a third of a nautical mile.[7]

This statement evidently related to the four ICBMs – the SS16, SS-17, SS-18, and SS-19 – that were flight tested in 1973-4 and were ready for operational deployment in early 1975. Since then, however, further refinements have been made to the guidance systems of these missiles, giving them CEPs, as at mid-1980, ranging from 0.14 nautical miles (850 feet) in the case of the Mod 1 version of the SS-19 to 0.24 nautical miles (1460 feet) in the case of the Mod 1 version of the SS-17.[8] Moreover, new guidance systems for the SS-18 and SS-19 ICBMs, first flight-tested in November 1977, have achieved a CEP of 0.1 nautical miles (600 feet); the retrofitting of these systems onto deployed missiles is currently proceeding and should be substantially complete by 1983-4 (i.e. some five years after the Improved NS-20 was incorporated on all the Minuteman III ICBMs).[9]

Increasing accuracy is also a feature of the American SLBM systems. The CEP of the Poseidon SLBM is currently about 1500 feet,[10] and the Improved Accuracy Program (TAP) should reduce this to about 1000 feet by the early 1980s. The Trident I (C-4) missile, which became operational in late 1979, can carry a full payload to a range of 4000 nautical miles while maintaining the equivalent of Poseidon accuracy, primarily through the use of a stellar sensor which takes a star sight during the post-boost phase of missile flight; the post-boost vehicle corrects its flight path based on this star sight.[11] There is, of course, no necessity for the Trident to be launched over its maximum range, and shorter flight paths produce correspondingly enhanced CEPs at the target; a Trident missile with a CEP of 1200 feet at maximum range would have a CEP of much less than 1000 feet if limited to the 2500 nautical mile range of the Poseidon SLBM. In any case, the Trident II (D-5) missile currently under development is expected to have a CEP of between 500 and 600 feet even at full range.[12] These low CEPs give US SLBMs a quite substantial counterforce capability, against all but the most hardened military targets.[13] The Soviet SS-N-8, which also has a stellar-inertial guidance system, reportedly has achieved CEPs of around 1200 feet, which also gives it some counterforce capability.[14]

A second aspect of the revolution in weapons technology is that the population systems of most types of weapons platforms – and weapons themselves – are being improved significantly. In particular there are several research and development efforts which appear to promise very much improved fuel, weight and space efficiencies. Engine developments include much more efficient solid propellant rocket booster motors, and relatively small but also highly efficient turbofan and turbojet engines – for use in, for example, strategic and tactical cruise missiles. While it has been the developments in modern guidance systems, and particularly of 'area correlation' techniques such as TERCOM (Terrain Contour Matching) which have most excited strategic analysts and planners with regard to cruise missiles, the strategic utility of these systems could not have been fully realised without the development of these small (about 130 lb) engines capable of powering the missiles over a range of 600 to 2000 miles.

Thirdly, with regard to explosive and warhead technology, not only has the destructive potential per given warhead volume and weight increased greatly in recent years, but a variety of new technology warheads have also been developed to meet specific requirements, primarily in the European theatre. On the American side, these include the B-61 variable-yield ('dial-a-yield') bomb; the three-option FUFO (Fuel-Fuzing Option) bomb with a device which enables the detonation of the free-fall bomb either in the air, on the ground, or by delayed action after it hits the ground, at the discretion of the bombardier; a series of so-called 'mininukes', or nuclear artillery shells for the US Army's 155 millimetre and 8 inch guns in Western Europe, which provide increased safety, security, range and quicker reaction time, as well as reduced collateral damage; and, of course, the enhanced radiation warhead (or so-called 'neutron bomb'). Though 'tactical' in deployment, these new developments have strategic implications. They (perhaps) lower the nuclear threshold, but improved battlefield effectiveness and enhanced flexibility provide a more continuous spread of capabilities across the lower end of the deterrence spectrum, thus leaving their implications for stability and deterrence rather ambiguous.

Some newly developed conventional explosives and warheads may also have strategic implications. Non-nuclear types of weapons such as fuel-air explosives (FAEs), improved cluster munitions for area targets, and hard structure munitions, when combined with highly-accurate long-range ballistic or cruise missiles, may be effective against a range of strategic targets which now require nuclear weapons for their destruction.

In addition to developments in purely weapons technologies, other technological advances are occurring in such areas as surveillance and early warning, target acquisition and target assessment, remote data processing power, survivable redundant command and control centres, etc., which also have important implications for strategic policies and national defence postures.

The Technological Developments in Strategic Perspective

Although these technological developments are quite disparate, they clearly tend in an identifiable direction – the enhancement of counterforce capabilities, damage-limiting strategies, strategic nuclear war-fighting options, and even first-strike possibilities, at least against the land-based components of strategic nuclear forces. This has been brought about principally through the very high single shot kill probabilities (SSKPs) achievable by ballistic missiles with the high accuracies described above. Even against targets hardened to withstand 2000 psi blast overpressure, each of the three Mark 12 re-entry vehicles currently deployed on the Minuteman III ICBMs, with a 170 Kt warhead and a CEP of 600 feet, would achieve a SSKP of about 60 per cent. However, the 900 Mark 12A re-entry vehicles to be deployed on 300 Minuteman III ICBMs, with 350 Kt warheads and CEPs of about 500 feet, will each have a SSKP against the same targets of about 90 per cent. With the deployment of the MX ICBM force of 200 missiles in 1986-9, the United States will achieve a full counterforce capability. Each MX missile is to carry 10 Mark 12A MIRV warheads, each with a yield of 350 Kt and a CEP of about 300 feet, giving each RV a

SSKP of greater than 95 per cent even against targets hardened to 3000 psi. The 200 MX missiles, each with a reliability of 75 per cent, would, therefore, be more than sufficient to destroy the whole Soviet ICBM force.

Further hardening of ICBM silos is incapable of significantly reducing the vulnerability of ICBMs to missiles with these high accuracies. Most of the US Minuteman ICBM force is now hardened to withstand about 2000 psi of blast overpressure.[15] The hardness of Soviet ICBMs, on the other hand, varies quite widely – the 60 SS-13s and the remaining 550 SS-11s that were deployed in the late 1960s and early 1970s are based in silos hardened to only about 300 psi; the major silo upgrading programme undertaken for the SS-17s, SS-18s and SS-19s in the mid-1970s produced resistance values of 2000-2500 psi; a further upgrading of some of these silos in the late 1970s increased this to between 3500 and 4500 psi; and a new silo design described as being about the size of the SS-17 and capable of withstanding 6,000 psi has recently been reported.[16] This probably represents the absolute limit to which missile silos can be hardened. Hardening to these very high resistances is extremely expensive (a 3000 psi silo costs more than $15m.) and is possible only in special geological environments. In any case, with CEPs of around 600 feet and below, hardening beyond 3000 psi reduces SSKPs by only a few percentage points.

Fixed-base ICBMs, then, are now obsolescent; by the mid-1980s, both the United States and the Soviet Union will have the capability, using only a part of their ICBM forces, to destroy a very substantial portion of the ICBMs of the other. The following tables describe the characteristics and hard-target capabilities of US and Soviet strategic nuclear forces as at the end of 1980 and as projected for 1990.

Table 1: US Strategic Nuclear Forces, December 1980

	Number of Delivery Vehicles	Throwweight (Thousand lb)	Number of warheads (n)	Yield per warhead (MT)	EMT per warhead	CEP (feet)	CMP per warhead	Total number of warheads (N)	Total MT	Total EMT	Total CMP
ICBMs:											
Minuteman II	450	1.6	1	1.2	1.1	1200	29.0	450	540	495	13050
Minuteman III (Mark 12)	450	2.4	3	.17	.3	600	50.2	1350	230	405	67770
Minuteman III (Mark 12A)	100	2.4	3	.35	.5	600	50.7	300	105	150	15210
Titan II	54	8.3	1	9.0	3.0	4250	6.1	54	486	162	329
SLBMs:											
Poseidon C-3	320	3.3	10	.05	.14	1500	5.0	3200	160	448	16000
Trident I C-4	176	2.9	8	.10	.21	1500	6.5	1408	140	296	9152
Bombers:											
B-52 D	75	9.6)	4 Gravity	1.00	1.0	600	102				
B-52 G	151	9.6)	bombs and	(bombs)	(bombs)	(bombs)	(bombs)	3792	1770	2124	160,528
B-52 H	90	9.6)	8 SRAM	0.20	.34	1200	12.5				
				(SRAM)	(SRAM)	(SRAM)					
FB-III	65	5	2 bombs & 4 SRAM	1.0/0.2	1/.34	600/1200	102/12.5	390	182	218	16,510
Totals	1,931							10,944	3613	4298	298,549

Notes: a. Equivalent Megatonnage (EMT) is the most meaningful index of destructive capability against 'soft' or 'area' targets (such as urban-industrial centres).

$EMT = Y^{\frac{2}{3}}$ where $Y \leqslant 1$ MT

$EMT = Y^{\frac{1}{2}}$ where $Y > 1$ MT

See Jeffrey T. Richelson, Evaluating the Strategic Balance', *American Journal of Political Science*, vol. 24, no. 4 (November 1980), pp. 797-800.

b. Counter-Military Potential (CMP) is the most meaningful index of destructive capability against 'hard' or 'point' targets (such as ICBM silos or underground command and control centres).

$CMP = \frac{Y^{\frac{2}{3}}}{(CEP)^2}$ where $Y \geqslant 0.2$ MT

$CMP = \frac{Y^{.4}}{(CEP)^2}$ where $Y < 0.2$ MT

See W.A. Barbieri, *Countermilitary Potential: A Measure of Strategic Offensive Force Capability* (The RAND Corporation, Santa Monica, R-1314-PR, December 1973), pp. 5-6.

Table 2: Soviet Strategic Nuclear Forces, December 1980

	Number of Delivery Vehicles	Throwweight (Thousand lb)	Number of warheads (n)	Yield per warhead (MT)	EMT per warhead	CEP (feet)	CMP per warhead	Total number of warheads (N)	Total MT	Total EMT	Total CMP
ICBMs:											
SS-11 Sego	550	2	1	1.0	1.0	5000	1.5	550	550	550	825
SS-13 Savage	60	1	1	.75	.83	6560	.7	60	45	50	42
SS-17 Mod. 1)	150	7	4	.75	.83	1460	14.3)	510	470	455	3,044
SS-17 Mod. 2)		7	1	3.6	1.9	1400	44.2)				
SS-18 Mod. 1)	308	16	1	24.0	4.9	1400	156.7)	2108	3160	1777	75,851
SS-18 Mod. 2)		16	8	.9	.93	1400	17.6)				
SS-18 Mod. 3)		16	1	20.	4.47	1155	204.1)				
SS-18 Mod. 4)		16	10	.5	.63	900	28.8)				
SS-19 Mod. 1)	330	8	6	.55	.67	850	34.3)	1830	1155	1268	63,813
SS-19 Mod. 2)		8	1	4.3	2.07	1275	69.1)				
SLBMs:											
SS-N-5 Serb.	21	n.a.	1	1.5	1.22	6000	1.3	21	31	26	27
SS-N-6 Sawfly Mod. 1)	469	1.6	1	.7	.79	3000	3.2)	1269	375	655	1,297
SS-N-6 Sawfly Mod. 2)		1.6	1	.65	.75	3000	3.1)				
SS-N-6 Sawfly Mod. 3)		1.6	3	.35	.5	4500	.9)				
SS-N-8	302	1.8	1	.8	.86	3000	3.5	302	242	260	1,057
SS-N-18 Mod. 1)	176	2.5	1	2.0	1.41	2000	14.7)	388	308	300	2,879
SS-N-18 Mod. 2)		2.5	3	.5	.63	2000	5.8)				
Bombers:											
Tu-95 Bear	113	8	4	1.0	1.0	3000	4.1	452	452	452	1,853
Mya-4 Bison	43	8	4	1.0	1.0	3000	4.1	172	172	172	705
Tu-22M Backfire	75	4	2	1.0	1.0	3000	4.1	150	150	150	615
Totals	2,597							7,812	7,110	6,115	152,008

Table 3: Projected US Strategic Nuclear Forces, 1990

	Number of Delivery Vehicles	Throwweight (Thousand lb)	Number of warheads	Yield per warhead	EMT per warhead	CEP (feet)	CMP per warhead	Total number of warheads (N)	Total MT	Total EMT	Total CMP
ICBMs:											
Minuteman II	450	1.6	1	1.2	1.1	900	51.6	450	540	594	23,220
Minuteman III (Mark 12A)	300	2.4	3	.35	.5	500	73.0	900	315	450	65,700
MX	200	9.2	10	.35	.5	300	207.0	2000	700	1000	414,000
SLBMs:											
Poseidon C-3	176	3.3	14	.05	.14	800	17.4	2464	123	345	42,874
Trident C-4	504	2.9	8	.10	.21	800	23.0	4032	403	847	92,736
Bombers:											
B-52 G	140	9.6	20 ALCM.	.2	.34	300	142.5	2800	560	952	399,000
B-52 H	90	9.6	4 gravity bombs and 8 SRAM	1/.2	1/.34	450/900	181.8/15.6	1080	505	605	87,912
FB-111	65	5	2 bombs and 4 SRAM	1/.2	1/134	450/900	181.8/15.6	390	182	218	27,690
Totals	1,925							14,116	3,327	5,011	1,153,132

Note: The total number of US strategic nuclear delivery vehicles projected for 1990 (viz.: 1,925) is 325 less than that permitted within the SALT constraints. This number could be made up with either a new single-warhead ballistic missile or a new non-cruise missile carrying bomber; however, the US has no current plans to produce either of these weapons systems.

Table 4: Projected Soviet Strategic Nuclear Forces, 1990

	Number of Delivery Vehicles	Throwweight (Thousand lb)	Number of warheads (n)	Yield per warhead (MT)	EMT per warhead	CEP (feet)	CMP per warhead	Total number of warheads (N)	Total MT	Total EMT	Total CMP
ICBMs:											
SS-18	308	16	10	.75	.83	500	121.4	3080	2310	2556	373,912
SS-19	362	8	6	.75	.83	500	121.4	2172	1629	1803	263,681
SS-17	150	7	4	.75	.83	500	121.4	600	450	498	72,840
New single-warhead ICBM	360	8	1	15.0	3.87	500	894.4	360	5400	1393	321,984
SLBMs:											
SS-N-18 Mod. 2	236	2.5	7	.2	.34	1200	8.8	1652	330	362	14,538
New MIRV	114	5	14	.2	.34	1000	12.6	2016	403	684	25,402
New single-warhead SLBM	570	2.5	1	2.0	1.41	1000	58.6	570	1140	840	33,402
Bombers:											
New Bomber	120	8	5	1.0	1.0	1200	25.6	600	600	600	15,360
Totals	2,250							11,050	12,262	8,900	1,121,119

Note: Assumes the Soviet Union continues to observe the SALT constraints.

In response to the development of these hard-target counterforce capabilities, both the Soviet Union and the United States are developing new basing modes for their ICBM forces. The SS-16 ICBM developed by the Soviet Union in the early 1970s was designed for land mobility, but although perhaps 60 SS-16s were produced the Soviets have agreed under the SALT II constraints not to deploy that missile. (However, more than 160 SS-20 intermediate-range ballistic missiles (IRBMs), which comprise two stages of the SS-16 are operational in a land-mobile mode.)

In the United States, the Carter Administration decided in September 1979 to begin full-scale engineering development of a system of multiple horizontal shelters for basing the new MX ICBM. Two hundred of these ICBMs are to be moved at random among 4600 shelters, spaced 6,000 feet apart so that no two shelters can be destroyed by any single Soviet warhead; the Soviet Union would need to launch 4,600 reliable and accurate warheads against these shelters in order to destroy the 200 MX missiles.*

Unfortunately, this is likely to be well within the capacity of the Soviet ICBM force by the late 1980s. Indeed, the viability of the MX basing system depends on some very specific assumptions about the character of the Soviet threat to the system, small changes in which could completely overwhelm it. The basic requirement underlying the MX force levels is that some 1000 ICBM warheads (i.e. 1000 re-entry vehicles, each with a significant hard-target capability) survive a Soviet first-strike attack. This number of surviving warheads would provide the capability to destroy most industrial targets in the Soviet Union or, alternatively, to attack a large portion of Soviet military targets. Carter Administration officials argued that if the Soviet Union wishes to deny this capability, she must fully disarm her own land-based ICBM forces, which would redefine the strategic balance in terms wholly of SLBMs and long-range bombers, where the US has a clear advantage. Given that the Soviets would use only the MIRVed portion of their ICBM force (i.e. the SS-17, SS-18 and SS-19 missiles) to attack US ICBMs; that a proportion of these MIRVs would be held in reserve (at least some of

*Editors note: This chapter was written before the announcement of the Reagan Administration's strategic programme. Some of the points raised by Dr Ball were crucial in the decision not to deploy MX in MPS. It should be noted, however, that a mobile option for MX has not been permanently ruled out by the Reagan Administration, and that the vulnerability of the MX in fixed silos, as now planned, is even greater than that suggested by Dr Ball.

which would be targeted on China); that two RVs would also be targeted on each of the 750 remaining Minuteman silos and another 300 RVs would be allocated for use against some 150 other counterforce targets in the US (such as ICBM command and control centres); that Soviet ICBM reliability is only about 75 per cent, and that the single-shot kill probability (SSKP) of each Soviet RV is greater than 95 per cent against shelters hardened to 600 psi, then the number of MX missiles surviving is 100, with the magic number of 1,000 RVs.[17] (See Table 5.)

Table 5: MX Survivability Calculus

Soviet MIRVed ICBM RVs	5852
Less 15% reserve, leaves	4974
Less 1500 RVs for use against Minuteman, leaves	3474
Less 300 RVs for use against other counterforce targets, leaves	3174
75% ICBM reliability, leaves	2380
97% SSKP	2300
4600 shelters less 2300 means that 2300 shelters with 100 MX ICBMs and 1000 RVs survive	

However, these assumptions underlying the MX survivability calculus are extremely rigid. There is some degree of trade-off possible among some of the variables – for example, one can assume that the proportion of RVs held in reserve is 20 per cent (1,170) rather than 15 per cent (880) and, on the other hand, that there is no targeting of the ICBM command and control centres (since the probability of destroying all the ICBM silos is sufficiently high anyway). However, if any of the principal assumptions are relaxed, the whole basing system loses its viability. For example, the Soviet Union could choose to use her MIRVed SLBM force against the shelters; this is expected to consist of 380 missiles each with some 7-14 warheads, each of which in turn is expected to have a significant counterforce capability. The Soviets could also decide to allocate one rather than two RVs to each of the 750 US silo-based ICBMs. By the late 1980s, SSKPs will be sufficiently high even against the 2,000 psi of the Minuteman silos that the second RV would only be necessary as insurance against the possible in-flight failure of the first; however, most sources of missile unreliability become apparent very soon after launch, so a second wave of RVs can be quite specific and relatively small. In any case, if Soviet ICBM reliability was itself to improve from the assumed 75 per cent up to 90 per cent then the number of surviving MX missiles would be almost halved.

Most importantly, however, the current MX basing mode possesses whatever viability it has only so long as the Soviets accept the SALT II constraints. If the Soviet Union was to abrogate the SALT agreements and deploy more ICBMs, or place more than 10 RVs on her MIRVed ICBMs, then the US ICBM force, including the MX system, would again be vulnerable to destruction by only a fraction of the Soviet ICBM force.[18]

Finally, developments in command, control and communication (C^3) technologies, including real-time satellite observation and warning, orders of magnitude improvements in data processing, and retargeting capabilities for both ICBM and SLBM systems, have greatly increased the capabilities of both the United States and the Soviet Union to conduct strategic nuclear strikes of a very limited and selective nature, or to fight an extended war of controlled sequential exchanges. Since 1961 at least, the United States has possessed a rather extensive capability for executing strategies of this sort, and declarations have periodically been made to this effect. The Soviet developments are much more recent but equally extensive. Satellites have been developed for defence communications, navigation, photographic reconnaissance, ocean surveillance, signals intelligence and early warning. An over-the-horizon (OTH) radar system which began test transmissions in July 1976 is being developed as an alternative long-range early warning system, and hardened command and control installations have reportedly proliferated, both in the Moscow area and within the ICBM fields. The achievement by the Soviet Union of a capacity to execute a flexible, controlled war-fighting strategy not only represents a significant change in Soviet strategic policy, but also signals a radical transformation in the Soviet-American strategic relationship. The operation of the MOLINK (Moscow-Washington link) satellite hot-line (with a Molniya ground station in Maryland and an Intelsat station near Moscow) to maintain continuous communications between the White House and the Kremlin even during a nuclear exchange is a recognition of this new relationship.

The Strategic Implications

The integration into national strategic policy of these enhanced capabilities for counterforce targeting and controlled nuclear exchanges has been most explicit in the case of the United States. In 1969, on the initiative of President Nixon and Henry Kissinger, various official studies were undertaken within the National Security Council and the Pentagon in an effort to devise strategies for limited and controlled nuclear exchanges. This work led to National Security Study Memorandum (NSSM)-169, approved by President Nixon in late 1973 and then, in turn, to National Security Decision Memorandum (NSDM)-242, signed by the President in January 1974.[19]

NSDM-242 contained three principal policy components. The one which engaged the most public debate was the re-emphasis on the targeting of a wide range of Soviet military forces and installations, from hardened command and control facilities and ICBM silos to airfields and Army camps.[20] The second element of NSDM-242 was the requirement for 'escalation control', whereby the National Command Authorities (NCA) should be provided with the ability to execute their selected options in a deliberate and controlled fashion throughout the progress of a strategic nuclear exchange. Thirdly, NSDM-242 introduced the notion of 'withholds' or 'non-targets', i.e. things that would be preserved from destruction. Some of these, such as 'population per se', have now been exempted absolutely from targeting; others, such as the centres of political leadership and control are exempted only for the purposes of intra-war deterrence and intra-war bargaining, and strategic reserve forces are to be maintained to allow their eventual destruction if necessary.

NSDM-242 also authorised the Secretary of Defense to promulgate the *Policy Guidance for the Employment of Nuclear Weapons* and the associated *Nuclear Weapons Employment Policy* (NUWEP), signed by Secretary Schlesinger on 4 April 1974. The NUWEP was developed through close military and civilian co-operation, and sets out the planning assumptions, attack options, targeting objectives and the damage levels needed to satisfy the political guidance. The concepts and objectives set out in NSDM-242 and NUWEP provided the framework for the development of new strategic nuclear war plans. The first Single Integrated Operational Plan (SIOP) prepared under the new guidance was SIOP-5, which was formally approved in December 1975 and took effect on 1 January 1976.[21]

Despite some initial expectation that the Administration of President Carter and Secretary Brown would move to change US policy back towards something more like Assured Destruction, the concepts and doctrines embodied in NSDM-242 and NUWEP have been essentially retained, and indeed further refined, through to the present day.[22]

These refinements were formally spelt out in Presidential Directive (PD)-59, signed by President Carter on 25 July 1980. The primary systems acquisition requirement identified in PD-59 was that the command, control, communication and intelligence (C^3 I) systems that control the SIOP forces should have greater endurance than the extant systems. PD-59 directed that more options should be added to the SIOP to give the strategic forces even greater targeting flexibility than they currently have. More specifically, it directed that there be relatively less emphasis accorded the destruction of Soviet economic and industrial targets and that greater attention be directed toward improving the effectiveness of US attacks against Soviet military targets. The most novel aspect of PD-59, however, is the requirement that greater attention be accorded the targeting of the Soviet military and political leadership and their command and control systems. The SIOP has always included some targets of this sort – for example, some 2000 of the 40,000 potential targets designated in SIOP-5D are leadership and control targets; however, it would not be unreasonable to expect that by the time SIOP-6 is authorised these targets would account for as much as 20 per cent of the Soviet target base – perhaps some 10,000 out of a likely total of 50,000 designated target installations.[23]

These developments in strategic doctrine were largely determined by the technological developments of the past decades. The ability to target a wide range of military installations, including those hardened to withstand thousands of pounds per square inch of blast overpressure, and the ability to conduct carefully delimited strikes was made possible by the development of accurate, multiple individually-targeted warheads and real-time surveillance, retargeting and communications capabilities. These technological developments not only created the strategic possibilities but in fact proved irresistible to the national security establishment. Despite the personal scepticism with which President Carter and Secretary Brown initially viewed the possibility of limited and controlled counterforce operations, neither was prepared to forego the possibilities that technology offered and thus to deny themselves the potential options and flexibility it allowed.

The allurement of the technology notwithstanding, however, there are in fact several major problems that would face any attempt at controlling selective and limited counterforce strikes. One problem, for example, is that even granted the precision with which intercontinental strategic nuclear delivery vehicles can now be targeted, the collateral casualties that would accompany any limited counterforce exchange are still very high. A comprehensive counterforce attack against the United States would involve strikes against the 1054 ICBM silos, two FBM submarine support bases (Bremerton and Charleston) and 46 SAC bomber bases.[24] Whereas the ICBM silos are generally located in relatively unpopulated areas, the two SSBN bases and many of the bomber bases are quite near major cities. Depending on the assumptions made about the scale and character of the Soviet strikes, US fatalities from such a comprehensive counterforce attack range from two to 20 million, with 14 million suggested as perhaps the most reasonable.[25] In the case of a comprehensive US counterforce attack against the Soviet Union, the targets would include nearly 1400 ICBM silos, three SSBN bases (Polyarnyy near Murmansk, Petropavlovsk on the Kamchatka Peninsula, and Vladivostok, the largest Soviet city in the Far East), 32 major air bases, and perhaps also the 700 IRBMs and MRBMs. Many of these are located in some of the most densely populated areas in the Soviet Union. (See figures 1-4 for maps of US and Soviet counterforce targets.) Twenty-two of the 32 major air bases, some three-quarters of the IRBM and MRBM sites, and more than half the 26 Soviet ICBM fields are located west of the Ural Mountains. Fatalities from a US counter-ICBM attack alone range from 3.7 to 27.7 million, depending principally on the level of fallout protection assumed; a full counterforce attack against bomber bases and FBM submarine support facilities as well as ICBM silos would obviously kill many more than this.[26] Four of the ICBM fields (Kozelsk, Teykovo, Kostroma and Yedrovo) are located sufficiently close to Moscow and distributed in such a way that the national capital would receive extensive fallout regardless of the prevailing wind direction.[27] Given casualties of this magnitude, and the particular Soviet difficulty of distinguishing a comprehensive counterforce attack from a more general military cum urban-industrial attack, the notion of limiting a nuclear exchange to supposedly surgical counterforce operations appears rather incredible.

A second problem pertains to the vulnerability of the strategic command, control and communication (C^3) infrastructure. Despite the fact that the US has spent some $25 billion on strategic C^3 over the past decade, much of it on means of enhancing the survivability of particularly critical C^3 systems, it remains the case that the C^3 infrastructure is more vulnerable and has less endurance than the strategic forces it is intended to support. Strategic C^3 systems are vulnerable to all the threats to which the forces could be subject plus a variety of additional ones. The strategic forces gain protection through hardening, proliferation, mobility and camouflage. Many C^3 systems, such as radar sites, VLF antennae and satellite sensor systems are necessarily relatively 'soft'; some C^3 elements, such as the National Command Authorities, cannot be proliferated; major command posts, satellite ground stations and communication nodes are generally fixed; and radar sites and communication stations are extremely difficult to camouflage because of their electronic emissions. C^3 systems are generally more vulnerable to the blast effects of nuclear weapons than are the strategic forces, and have various peculiar vulnerabilities as well – susceptibility to electromagnetic pulse, electronic jamming, deception, etc.

Figure 1: Counterforce Targets in the United States

9 ICBM sites: 1. Malstrom – 200 Minuteman; 2. Ellsworth – 150 Minuteman; 3. Minot – 150 Minuteman; 4. Grand Forks – 150 Minuteman; 5. Warren – 200 Minuteman; 6. Whiteman – 150 Minuteman; 7. Davis-Monthan – 18 Titan II; 8. Little Rock – 1 Titan II; 9. McConnell – 18 Titan II.

The vulnerabilities of such critical elements of the strategic C^3 architecture as the National Command Authorities (NCA) themselves; the airborne C^3 systems which are relied upon for continuity of command and control in a nuclear environment; the satellite systems used for communications, early-warning, photographic reconnaissance and signals intelligence; the 'hot line' which would be required for communication and negotiation between the adversary leaderships; and the communication systems for the FBM submarines, which at least in the US case carry half the warheads of the strategic nuclear forces – these impose quite debilitating physical constraints on the situations in which escalation might be controlled, the time period over which control might be maintained, and the proportion of the SIOP forces that could be employed in a controlled fashion. The boundary of control in any militarily significant exchange (as compared to demonstration strikes) is unlikely to lie beyond either a few days or a few tens of detonations.

Figure 2: Soviet ICBM Deployment

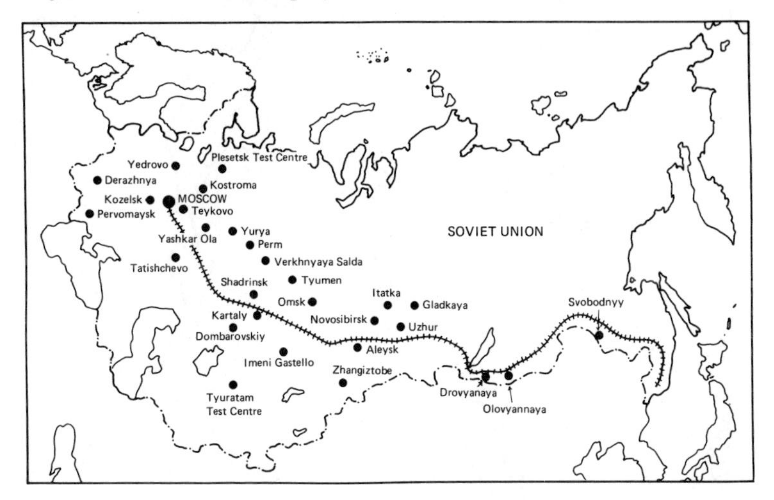

The technological developments of the past decade were necessary to the improvement of counterforce capabilities, damage-limiting strategies and escalation control policies. However, they are not sufficient for the successful implementation of counterforce and controlled response doctrines. Other factors which must be taken into account, and which generally augur against the operational success of such doctrines include the inability to conduct extensive counterforce operations without causing significant collateral casualties and damage, the inherent vulnerabilities of strategic C^3 systems, a number of residual technical uncertainties (such as missile performances in an operational environment), the question of whether the Soviet Union would co-operate in maintaining restraints on an exchange or would act massively against the whole US target structure, and all the political and emotional pressures that would accompany even the 'limited' use of nuclear weapons. The great danger is that as well as succumbing to the strategic promise of the technological developments of the past decade, the United States and the Soviet Union might also lose sight of these other realities.

Figure 3: Major Soviet Air bases

Figure 4: Titan II Bases and Population Co-location

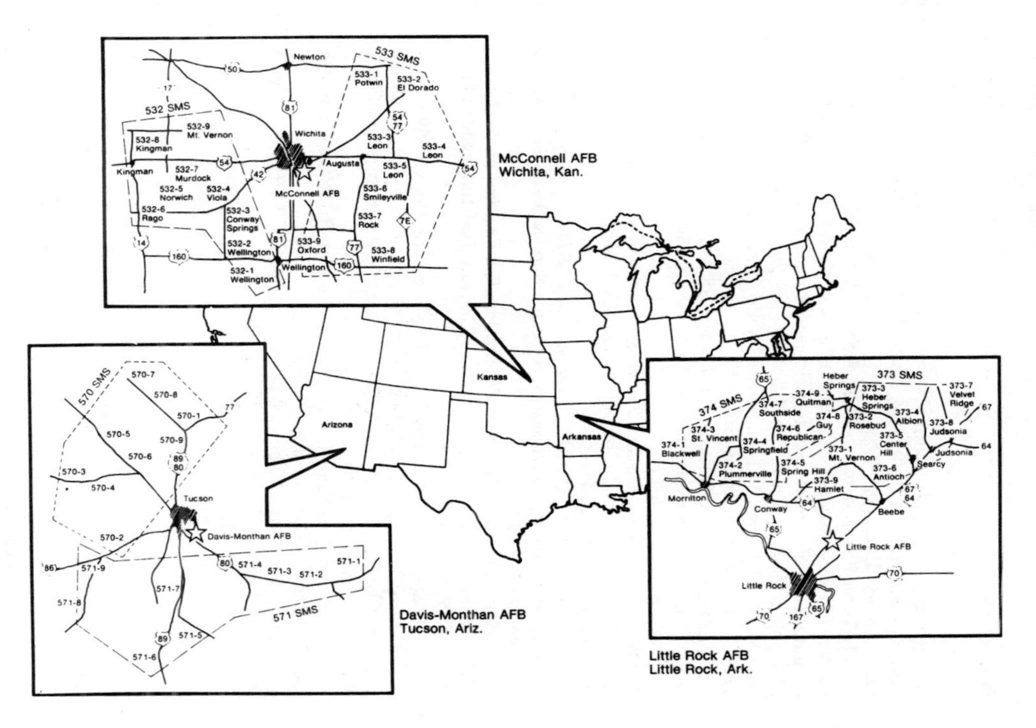

Notes

1. See the two-part technical survey of particle beam and high-energy laser developments in *Aviation Week & Space Technology*, 28 July 1980, pp. 33-66, and 4 August 1980, pp. 44-68.

2. According to *Aviation Week & Space Technology*, 6 August 1979, p. 13, the Soviet Union may have already tested a small, space-based hydrogen-fluoride chemical laser. For references to a similar US system, reportedly tested in 1977, see Robert Lindsey, *The Falcon and the Snowman* (Simon & Schuster, New York, 1979), pp. 111-2; and Joe Trento, 'Cuba Crisis Tied to US Laser Gun', *The News Journal*, 8 September 1979.

3. *Aviation Week & Space Technology*, 28 April 1980, p. 20.

4. 'Spacecraft Survivability Sought', *Aviation Week & Space Technology*, 16 June 1980, pp. 260-1.

5. *Aviation Week & Space Technology*, 28 August 1978, p. 18; and testimony of Paul H. Nitze, Senate Foreign Relations Committee, *The SALT II Treaty* (Part 1), July 1979, p. 458.

6. A.A. Tinajero, *The MX Intercontinental Ballistic Missile Program*, (Library of Congress, Congressional Research Service, Issue Brief No. IB77080, 29 June 1977), p. 19.

7. Senate Foreign Relations Committee, *Briefing on Counterforce Attacks* (Secret Hearing held on 11 September 1974; sanitised and made public on 10 January 1975), p. 10.

8. *Aviation Week & Space Technology*, 16 June 1980, p. 69.

9. Jim Klurfeld, 'The MX Debate', *Long Island Newsday*, 3 February 1980, p. 5; and Clarence A. Robinson, 'Soviets Boost ICBM Accuracy', *Aviation Week & Space Technology*, 3 April 1978, pp. 14-16.

10. Testimony of Paul H. Nitze, Senate Foreign Relations Committee, *The SALT II Treaty* (Part 1), July 1979, p. 458.

11. R.T. Pretty (ed.), *Jane's Weapon Systems 1979-80* (Macdonald & Jane's Publishers Limited, London, 1979), pp. 21-2.

12. *Air Force Magazine*, June 1980, p. 17. *Aviation Week & Space Technology* has even reported (8 December 1980, p. 11) that the CEP of the Trident II SLBM is expected to be as good as 400 feet.

13. See Desmond J. Ball, 'The Counterforce Potential of American SLBM Systems', *Journal of Peace Research*, vol. XIV, no. 1 (1977), pp. 23-40.

14. Doug Richardson, 'Soviet Strategic Nuclear Rockets Guide', *Flight International*, 11 December 1976, p. 1733.

15. Bruce W. Bennett, *How to Assess the Survivability of US ICBMs* (The RAND Corporation, Santa Monica, R-2577-FF, June 1980), p. 5. However, the 150 Minuteman II silos at Ellsworth Air Force Base, South Dakota, are hardened to somewhat less than 2000 psi. See testimony of General Slay, Senate Appropriations Committee, *Department of Defense Appropriations for Fiscal Year 1979*, (US GPO, Washington, 1978) (Part 4), p. 985.

16. Colin S. Gray, *The Future of Land-Based Missile Forces*, Adelphi Papers no. 140 (IISS, London, Winter 1977), pp. 12, 17; *Aviation Week & Space Technology*, 16 June 1980, p. 67; and *Aviation Week & Space Technology*, 3 November 1980, p. 28.

17. See Desmond Ball, 'The MX Basing Decision', *Survival*, vol. XXII, no. 2 (March-April 1980), pp. 58-65.

18. The 1979 National Intelligence Estimate on Soviet strategic offensive forces, NIE 11-8-79, noted that as many as 30 warheads could be deployed on each SS-18 ICBM force and that without the SALT constraints the Soviet

MIRVed ICBM force alone could amount to some 14,000 warheads by 1989. See Michael Getler and Robert G. Kaiser, 'Intelligence Estimate Said to Show Need for SALT', *Washington Post*, 31 January 1980, p. 1.

19. House Appropriations Committee, *Department of Defense Appropriations for 1975*, (US GPO, Washington, 1974), Part 1, p. 499; House Armed Services Committee, *Hearings on Military Posture and H.R. 1872*, (US GPO, Washington, 1979), Book 1 of Part 3, pp. 6-26.

20. See, for example, Senate Foreign Relations Committee, *US-USSR Strategic Policies* (Top Secret hearing held on 4 March 1974; sanitised and made public on 4 April 1974), pp. 18-19.

21. Senate Armed Services Committee, *Fiscal Year 1977 Authorization for Military Procurement, Research & Development, & Active Duty, Selected Reserve & Civilian Personnel Strengths* (US GPO, Washington, 1976), Part II, p. 6422; House Appropriations Committee, *Department of Defense Appropriations for 1977* (US GPO, Washington, 1976), Part 8, p. 30; House Appropriations Committee, *Department of Defense Appropriations for 1980* (US GPO, Washington, 1979), Part 3, p. 878; and House Armed Services Committee, *Hearings on Military Posture & H.R. 1872* (US GPO, Washington, 1979), Book 1 of Part 3, pp. 6-26.

22. See Desmond Ball, *Developments in US Strategic Nuclear Policy Under the Carter Administration*, ACIS Working Paper No. 21 (Center for International and Strategic Affairs, UCLA, Los Angeles, February 1980).

23. See Desmond Ball, 'PD-59: A Strategic Critique', *Public Interest Report*, (Newsletter of the Federation of American Scientists), vol. 33, no. 8 (October 1980), pp. 4-5.

24. Senate Foreign Relations Committee, *Briefing on Counterforce Attacks*, pp. 10-18.

25. Office of Technology Assessment, US Congress, *The Effects of Nuclear War* (Croom Helm, London, 1980), pp. 81-6.

26. Ibid., p. 91.

27. Desmond Ball, 'Research Note: Soviet ICBM Deployment', *Survival*, vol. XXII, no. 4 (July-August 1980), pp. 167-70.

7 TECHNOLOGY, DETERRENCE AND THE NATO ALLIANCE

Lothar Ruehl

The Present Situation

In order to set the context for a discussion of recent and imminent changes in theatre nuclear technology, and the implications they suggest for deterrence stability and escalation dominance, it is useful to outline briefly the general structural characteristics and trends which form the more particular Eurostrategic environment.

First, it is clear that while both the Warsaw Pact and the NATO Alliance depend on the linkage of defence and deterrence through the medium of escalation, there are critical differences in the substance of this relationship as between East and West. The major nuclear power of the East – the Soviet Union – has a monopoly of nuclear arms, does not deploy long-range nuclear systems beyond its national borders, and has not distributed nuclear systems amongst its allies in a manner or magnitude similar to that of the United States in NATO. The consequences of this basic structural asymmetry are major for Alliance politics, strategy and arms control policies:

1. the Soviet Union can propose withdrawals of nuclear systems from Europe without being hampered by Alliance arrangements;
2. in an operational and political sense, the Soviet Union exercises much tighter control over nuclear forces than does the United States *vis-à-vis* France and Britain;
3. for the Soviet Union, nuclear strategy is basically a prerogative of the USSR, while in NATO it is subject to Alliance discussions, often leading to dissent, or even withdrawal, as with France in 1966-7. NATO is basically a nuclear coalition between a nuclear Superpower, two medium nuclear powers, and a dozen countries not possessing nuclear weapons, but with some of them operating US weapons under US custody and command and control.

Secondly, while the significance of conventional forces has been recognised by NATO, this has been primarily in an escalation context: i.e. as a tangible base-line for the credible articulation of a strategy of deterrence. At this level, significant improvements have taken place in the conventional posture of both NATO and the Warsaw Pact since the mid-1960s. Nevertheless, for NATO the result is still inferior to the 'stalwart defence' called for by US Secretary of Defense James Schlesinger in 1973-4. Moreover, it is difficult, given current military-industrial capabilities and political realities, to see how it could be otherwise. The Soviet industrial base and tactical and strategic reliance on superior firepower at all levels, make it improbable that NATO could ever make up for Soviet capabilities merely through counter-deployments of conventional forces.

Precision-guidance technology for high accuracy weapons offers some promise for offsetting superior tank, aircraft and artillery numbers. But even if such systems were deployed in sufficient numbers, and appropriate logistical, tactical, and command and control adjustments were made, there are still real limitations to the potential of first- and second-generation PGM systems for overcoming Eastern offensive capabilities. This arises from the critical dependence of these technologies, as currently deployed, on clear lines-of-sight, and often on daytime operation. This is not to imply that they are useless, or weigh lightly in the balance; PGMs and associated data processing and C^3 developments can significantly upgrade defensive options. But they can also serve offensive operations, and have been deployed in such a mode by the Soviet Union since the early 1970s. One case in point is the use of such light anti-air and anti-tank guided weapons by the Egyptian infantry for the offensive across the Suez Canal in the October 1973 War.

Thirdly, while conventional PGM and low yield/high accuracy nuclear systems have been regarded in the West as solutions to the costs, attrition rates and stockpiling requirements of classical conventional arms, they have been coupled in the East with tandem improvements in chemical and nuclear war-fighting forces. In the Warsaw Pact, technology has yet to be made a *substitute* for quantity but is seen as an enhancement to the traditional advantages deriving from larger forces.

It is this double effort which has complicated NATO plans to use nuclear arms in a support role for compensation against superior conventional attacks. Much points to the probability of swift escalation of any military conflict to nuclear war on the assumption that this is part of the main body of Soviet military doctrine.

Fourthly, for purposes of rationalisation and enhancement of escalatory control and refinement, command and control over nuclear and conventional forces has been upgraded within NATO in recent years. Unreliable, inaccurate, vulnerable, or excessively capable nuclear weapons have been progressively eliminated, beginning in the early 1960s under the aegis of Robert McNamara. The determining factors here have been available technology, improved organisation for command and control, and a preoccupation with reducing the risks associated with the premature or disproportionate use of nuclear arms. Nevertheless, NATO's nuclear position in Europe has remained essentially unchanged, and inappropriate for the adequate operational implementation of 'flexible response'. American decision-makers such as Kissinger and Schlesinger have warned against both a purely symbolic attitude towards them, and an attitude which would rely on them for the security of Europe. Symbolism is inadequate, and sole reliance is both dangerous and incredible. This theme was repeated by Henry Kissinger in his famous 1979 Brussels *démarche*, with regard to the whole question of the American nuclear posture.

Finally, the most significant and active area for real change in European security has been in long-range theatre nuclear forces – LRTNF or 'Eurostrategic' arms. Historically, since the withdrawal of medium-range nuclear systems from Turkey and Europe in the early 1960s by the United States, 'external forces' – ICBMs, SLBMs, and strategic bombers – have been relied upon to counter the Soviet advantage in medium- and intermediate-range ballistic missiles (M/IRBMs). The question until recently has thus been: would these continental weapons be upgraded to significant new levels when replaced by more modern technology? The SS-20 and Backfire deployments since 1977 have answered this question, and while American forward based systems – the Poseidon SSBNs and FB-111 medium-range bombers – could and can be regarded as providing *some* countervailing power to these Soviet systems, the main Soviet effort in the 300-1000km range has gone largely unanswered.

Thus, Western deterrence below global and central intercontinental strategic arms was undercut, leading concerned European allies, in particular West Germany, to plead both for appropriate counterdeployments in order to maintain the escalatory fabric, and meaningful and stabilising bilateral arms control negotiations at the European-nuclear level. (At lower levels, the enhanced radiation weapon (ERW) and upgraded battlefield systems, such as the Lance, have been discussed with decidedly mixed success, thus far. The Lance deployment can be looked upon as an operational improvement in terms of flexibility, mobility, survivability and target acquisition. But Lance missile systems do not replace the older Honest John missiles one for one; the overall reduction in numbers of launchers is about one for three, or 66 per cent.)

The political process within NATO between 1977 and 1979 resulted in a decision in December 1979 to upgrade Alliance LRTNF assets, in order to provide improved operational flexibility, survivability, accuracy, penetrability, and command and control. It is these developments, and the Soviet deployments to which they are designed to respond – the SS-20, Backfire, SU-19, the various SRBM systems of various ranged from the SS-21 to the SS-24, which give rise to the most significant questions with regard to the strategic and operational conditions for deterrence – and defence – in the years ahead.

Foreseeable Challenges and Solutions

The central strategic significance of recent improvements in Soviet long-range theatre forces – often called 'continental strategic arms' – lies in the perceived capacity such systems give the Soviet Union with regard to the matrix of problems generally referred to as 'theatre nuclear war/decoupling/and threat leverage.' In this connection, it is argued that the qualitative aspects of such weapons as the SS-20, Backfire, Fencer (SU-19) combat aircraft, and the SS 21-24 tactical missiles, offer additional offensive options – and enhance threat potential – against Western Europe. Hence they are often seen as directed towards the development of a capacity to dominate in a political-military crisis arising in the central theatre. However, while this argument has some merit, and the weapons systems to which it refers are far from trivial in their consequences, the real question is not whether such systems are 'improvements', or whether they increase the theatre nuclear asymmetries in central Europe. Rather the critical issue is whether the deployment of these weapons actually *destablises* the existing force relationship, leading to a balance of options in a crisis situation which would lead the Soviet Union to accept risks which it would otherwise not be prepared to accept, in effect to use these enhanced capabilities to dominate, coerce, conquer or destroy.

In technical terms, the argument is often made that these new Soviet systems, in their qualitative and quantitative characteristics, are capable – in a new sense – of interdicting existing NATO LRTNF assets through pre-emption, hence depriving NATO of the means of nuclear escalation at the moment of greatest need. This surgical threat to the escalatory repertoire of the Western Alliance would, in such a case, undermine, and render inoperable, NATO's strategy of flexible response.

It is in response to these perceived strategic developments and plausible scenarios that the modernisation of NATO's LRTNF systems has been directed, in order to provide not merely greater overall destruction capability – indeed it may reduce it – but to enhance those qualities which ensure survivability: mobility, shorter reaction time, improved accuracy, and more sophisticated target acquisition. Pershing II missiles (range: 1800 km) and GLCMS (range: 2500 km) have been selected for this purpose, with an overall planned deployment in Europe of 572 warheads (108 Pershing IIs; 464 GLCM). Both are equipped with 'near terminal guidance', according to US Department of Defense statements made at the November 1979 NPG meeting. The strategic significance which can be assigned to these forces – on both sides – depends on assumptions made with regard both to operational intent and plausible psychological effect. If one assumes that, for example, the SS-20 force is – or will be – capable of effectively neutralising the current 400-600 land-based LRTNF installations, this must be based on the quantities and qualities of the SS-20 force deployed: its range, accuracy, and mobility characteristics. On the other hand, to assert merely that, on the bases of range and destructive potential, a greater countervalue threat can be posed against Western Europe would be to say considerably less of *strategic* significance. An increased countervalue threat is, in and of itself, of minor *qualitative* consequence to the dynamics of nuclear escalation in Europe, although it would reinforce current patterns substantially. The critical factor is the yield-accuracy combination for selective and limited strike options.

Mobility is a crucial issue in both cases. For by reducing system vulnerability, the dangers of pre-emption would be dramatically reduced; this in turn would allow the firing of SS-20 missiles in sequential echelons. By fractionating an LRTNF force into consecutive launch salvoes, to be used over a long period – by nuclear war standards – of several days, the theatre element of the overall deterrent would be successfully divorced both from general total use at that level and from the central nuclear relationship. It is for *this* reason that the 750 Soviet IR/MRBMs of the SS-4/5 class – about 500 of which were deployed against Europe – could be regarded with relative equanimity in NATO circles. The characteristics of the replacement force are such as to thaw the previously frozen all-out attack posture. As long as such a restrictive employment requirement exists, a Soviet theatre nuclear strike could not be contemplated except as part of an overall ballistic missile attack against both Europe and the United States. The vulnerability of the SS-4/5 missile launchers to pre-emption would mean that if these weapons were not launched simultaneously in the context of a global exchange, many of them could be destroyed on the ground after the first IRBMs were launched. Moreover, this pre-launch vulnerability, combined with their indiscriminate and massive size (1-2 MT), meant that any IRBM strike would be very likely perceived as a total provocation. Thus the existence of a Eurostrategic imbalance in the past was tolerable in a *strategic* sense, given the operational characteristics and escalation dynamics such a Soviet force implied. For example, even if US retaliation was not forthcoming, it is very likely that such an attack by Soviet SS-4/5 assets would be swiftly followed by retaliation from French SSBS against the 30 biggest Soviet cities within range. The risks of this retaliation would engender great caution, and restrictive operational limitation to the use, or threatened use, of pre-SS-20 Soviet IRBM forces.

At the current juncture, however, it remains to be asked if the SS-20 has changed this fundamental strategic reality. To some extent it has. The missile's characteristics – a range of more than 4,000km, capacity for re-load, mobility, accuracy and MIRVed capability (3 x 200 kt) – render it ideal for selective attacks against more or less isolated military targets. If such a capability existed – or was imminent – a re-thinking of the bases for French and NATO escalatory postures would be required. The emergence of such a situation can be ascertained in the near future as logistical and testing patterns render operational capabilities and intentions clearer.

The unanswered question, then, remains the targeting criteria adopted by the Soviet Union with its SS-20 force. In so far as this capability is used to cover NATO rear areas – ports, supply depots, airbases etc. – then the level of consequent damage would be such as is traditionally considered sufficient to provoke swift escalation to strategic levels of general war. If this were the case as regards the SS-20, and shorter-range systems were used in a similar manner closer to their launch points, then the Soviet Union would not be contemplating the sort of 'limited' nuclear war which would block the escalatory link, serving to decouple NATO central and theatre nuclear forces.

Nevertheless, it *is* clear that whatever target systems are selected for the SS-20, its qualitative superiority over the SS-4/5 indicates a major change in the nature and extent of the IRBM threat to the entire NATO-European area. The deployment of weapon systems in Eastern

Russia or Siberia which can strike Western Europe without being exposed to NATO's European-based arms and, to a large extent, to *any* arms at all, constitutes a basic change in the East-West balance of deterrence in Europe. This both explains the deployment *and* the qualitatively different Western reaction to the deployment of the SS-20.

The argument, then, need not depend on expressed or imputed Soviet intentions, nor on the numbers or yield of the systems deployed, but rather on the specific technical and operational characteristics of these systems and the logical consequences of such weapons for conflict in Europe. It may well be that the SS-20 is merely a follow-on to older IRBMs, a product of 'natural' modernisation and technological progress – as Backfire is to the TU-16, but the truth of this fact, which is self-evident, does not detract from the strategic implications just outlined.

On the Western side, the cruise missile, Pershing II MRBM, and enhanced radiation weapons, have been singled out by the Eastern bloc as 'destabilising'. The arguments here are the same:

1. these weapons change the set-piece battle of nuclear arms, and hence the conditions of conflict, through their characteristics;
2. in offering additional options, they constitute additional challenges to the East;
3. they will provoke counter-deployments which will leave the balance similar to that which existed before, although at a higher level of destructive power and destabilising sophistication.

Thus the mutual perceptions of the destabilising character of new technologies seem to come full circle. But the geopolitical asymmetries of the current imbalance indicate that this assumption of apparent uselessness – and even danger – from new Western theatre deployments is misguided. For through new technology *and* geopolitical position, the Soviet Union can cover Western Europe with various categories of nuclear and conventional arms, while NATO – in Western Europe – has only marginal capabilities to pose a counter-threat to the Soviet Union and Warsaw Pact forces behind the front lines.

Historically, NATO has had to rely on three mechanisms of extended deterrence – other than threat of collective suicide: central US strategic systems used in a selective manner against particular Soviet targets; forward-based systems to strike at targets in Eastern Europe and the Western and South-Western USSR; and the combination of tactical nuclear weapons and conventional forces for battlefield use. It remains to be seen, therefore, how NATO strategy can be maintained and rationalised in a coherent posture with the new weapons available, and technological challenges from the East.

In this sense, in order to respond to Soviet deployments, NATO LRTNF capable of reaching Soviet territory must be deployed for limited and selective strikes in order to serve 'flexible response'. Moreover, targets must be selected on the basis of the interdiction of Soviet war-fighting options, rather than relying on an obsolete and ineffectual deterrent threat of massive retaliation against Soviet targets in response to limited strikes against West European military targets.

There is an additional, and perhaps contradictory, aspect to Western reasoning with regard to Eurostrategic deployments. These forces are designed both to serve useful defensive purposes in the process of war-fighting, and to implement flexible response through the threat and reality of escalation. This in turn requires that Soviet forces and nuclear assets be attacked fairly early in the game. (Presumably, the range of around 2000 km in the context of planned GRCM deployments is designed to do this.) However, the limited number of warheads to be deployed in central Europe will produce a highly time-sensitive force, and hence a rather low nuclear threat level for optimal use. If, therefore, technically and operationally optimal use is to be made of these modern Western LRTNF, this would indeed change the prospects of the East in Europe through a radically changed nuclear dynamic.

Soviet statements have argued that ballistic missiles in Western Europe would reduce pre-attack warning for parts of the USSR to about six minutes, and hence would represent an entirely new type of threat to the Soviet Union. Aside from the technical point that 'warning time' should not be equated with time between launch and impact, it is none the less clear that this threat would put the Western USSR in exactly the same position as Western Europe is – and has been – *vis-à-vis* the USSR. Thus, for the first time in more than a decade, technology offers the potential to compensate for geopolitical asymetries at a critical level in the deterrence/escalation balance.

8 NATO AND LONG-RANGE THEATRE NUCLEAR WEAPONS: BACKGROUND AND RATIONALE

Paul Buteux

In December 1979, at a special meeting of foreign and defence ministers, it was agreed that NATO should modernise its long-range theatre nuclear forces (LRTNF). This was to be done by deploying 108 extended range Pershing II launchers in Germany in place of existing Pershing Ia missiles, and also by deploying 464 ground launched cruise missiles (GLCMs) with a range of 2,500 kilometres in a number of allied countries. All these missiles would have single warheads, and together would replace 572 warheads already deployed. In addition, 1,000 other warheads were also to be withdrawn as part of the modernisation package. Finally, the decision on modernisation was linked to an arms control proposal that called on the Russians to begin negotiations with the United States for mutual limitations on the long-range nuclear systems that each had deployed in the European theatre.[1]

The decision had a wide range of political, strategic and arms control implications, and served to underline the impact on the Alliance of substantial changes in the strategic environment affecting allied security. The substance of these changes can be summed up by the phrase 'strategic parity'; the consequences for the military balance in Europe of parity between the strategic forces of the United States and the Soviet Union had become of increasing concern to the European allies. As the allies approached the modernisation decision many doubts and hesitations emerged and securing agreement seemed to be a difficult task. The decision on modernisation was controversial and a difficult one for the Alliance to reach. Nevertheless, a consensus was achieved sufficient for the first steps to be taken towards modernisation of the long-range nuclear forces available to the Alliance in the European theatre.

The decision to modernise the Alliance's LRTNF forces represented one attempt to deal with the problems created for European security by the fact of strategic parity. In particular, the decision can be seen as an attempt to reinforce the credibility of the American strategic commitment to the security of its allies in Europe. The declared purpose of American strategic forces is not only to deter a Soviet attack on North America, but also to deter any threatened attack on the allies of the United States in Western Europe. This involves the United States in a posture of extended deterrence, and it is the purpose of this chapter to examine in what ways the deployment of a force of modernised LRTNF might reinforce the credibility and political effectiveness of that extended deterrent commitment.

Ever since the Soviet Union obtained in the early 1960s a significant strategic retaliatory capability against the United States, the credibility of the American strategic commitment to Western Europe has been challenged. It has been argued that given the ability of the Soviet Union to retaliate, it is unlikely that the United States would initiate a strategic nuclear exchange with the Soviet Union except in terms of a direct and immediate threat to itself. The United States would wish to respond to any assault on its European allies at much lower levels of violence than would be implied by strategic strikes directly against Soviet territory. The development of the strategic concept of flexible response was an important way in which the United States sought to give continued credibility to a nuclear commitment to its allies in circumstances in which American territory was becoming increasingly vulnerable to Soviet retaliation. The adoption by the Alliance in 1967 of flexible response as the strategic concept on which its military planning would be based, represented a recognition on the part of the Alliance that the credibility of the extended deterrent depended on the ability of the Americans to limit the risks involved in their commitment to European security.

Put briefly, through the concept of flexible response the Alliance seeks to maintain the military means to respond to any Soviet or Warsaw Pact action at a level appropriate to the threat offered. Should the Alliance fail to achieve the desired effect at any particular level of military response, then the strategy holds out the option of escalating to higher levels of response. This includes, if need be, the option to initiate the use of nuclear weapons. The argument is that if the Alliance possesses the ability to respond to any aggression at a level commensurate with the assault, this, together with the threat of escalation, should persuade the adversary of the great risks and costs of continued violence. The overall credibility of this strategy is held to be dependent on the ability of the allies to provide a range of military options from conventional to nuclear and to link them together in an escalatory chain that includes the strategic forces of the United States.

To a significant extent, the political acceptability of flexible response to the NATO allies has always depended on the assumption that in the event of escalation at the nuclear level, the strategic and, hence, bargaining advantage would with NATO. Implicit in the threat to escalate through the nuclear threshold has been the belief that the Alliance possessed strategic superiority. Once parity at the level of strategic weapons had been conceded however, and this had been formally recognised in the SALT process, then any escalation advantage enjoyed by the Alliance necessarily would have to rest on an assumed superiority at the level of theatre nuclear forces. But at this level also, the Soviet Union and her allies in the Warsaw Pact have established in recent years at least overall parity in theatre nuclear forces, and in terms of 'static' criteria have a distinct advantage in long-range systems.[2]

In particular, two new Soviet long-range theatre systems have given rise to a great deal of Alliance concern. These are the SS-20 mobile medium-range ballistic missile fitted with multiple warheads and the Backfire bomber. When fully deployed, these new weapons will give the Soviet Union a greatly improved nuclear counterforce capability over the whole of Western Europe, in addition to placing most major Western European cities at risk. Compared with the earlier generation of what have been termed 'Eurostrategic' forces available to the Soviet Union, its modernised LRTNF offer much greater flexibility and a wider variety of strategic options in the European theatre. (The long-range theatre nuclear forces that are being discussed here have been termed 'Eurostrategic' because they fall outside the accepted SALT definitions of what constitutes a strategic weapon; none the less, were they to be used in the European theatre, the scale of effect would be comparable to that which would result from a strategic exchange involving the central forces of the two major nuclear powers.)

Admittedly, the significance of theatre nuclear parity, the extent to which numerical advantage confers any meaningful strategic and political advantage, and the implications of asymmetries in the force postures of the two sides in Europe, are all matters of some controversy. However, there is a widely shared recognition that the Alliance no longer possesses any clear nuclear superiority at either the theatre or strategic level. This state of affairs poses challenges to the viability of the strategy of flexible response and therefore to the credibility of the extended deterrent posture based on it.

This raises the specific question of what contribution the decision to deploy a force of modern LRTNF might make to enhancing the credibility of flexible response. Prior to LRTNF becoming the central element in the Alliance debate on theatre nuclear weapon modernisation, attention had been directed primarily to battlefield weapons, and to the escalation and crisis control implications of improved command, control and communications (C^3), and to the need for ensuring greater invulnerability for theatre nuclear forces. The modernisation of LRTNF, however, places emphasis on the military and deterrent function of interdiction forces, and on the extent to which there is a NATO requirement for long-range, theatre counterforce capabilities. This necessitates, in turn, a more direct consideration of where LRTNF fit into a pattern of deterrence which is characterised ideally by its graduated and flexible character. To what extent do long-range theatre nuclear forces increase the ability of NATO to respond effectively and appropriately to any attack either threatened or launched?

The theatre nuclear posture of NATO has always been characterised by ambiguity as to the actual war-fighting role envisioned for the Alliance's theatre nuclear forces. It is not that there are no military plans for the use of these weapons; but, rather, that their relationship to the perennial question of 'deterrence or defence?' is not clear. The extent to which the Alliance should rely for its security on deterrence by threat of punishment, rather than by threat of denial (the familiar distinction between deterrence and defence), has been a matter of constant debate since nuclear weapons first became part of NATO military planning in the 1950s. Moreover, judgements as to the value of a modernised long-range theatre nuclear force will be affected by the position with respect to this debate, and with estimates of the impact of such modernised forces on the problems of deterrence and defence. For, after all, the requirements for a reprisal force of long-range theatre nuclear weapons are not necessarily the same as a force designed to affect the outcome of a nuclear battle.

If the object is to deter by threat of punishment, and presumably in a graduated scheme of deterrence the object is to 'make the punishment fit the crime', the extent to which the Alliance needs war-fighting capabilities may be questioned. In addition, different judgements as to the need for a war-fighting capacity may be reached with respect to different levels of nuclear conflict. For example, at the battlefield level, it would seem that the political commitment to forward defence in Germany requires a military posture that holds out some prospect of limiting the amount of destruction to the defenders' own territory. In so far as this involves the use of nuclear weapons to affect the immediate outcome of the battle, then a use which attempts to limit collateral damage, and which seeks to halt the attack and disrupt the enemy's forward echelon forces, would seem to be required. In an intense, fast-moving battle in Central Europe, the impact of long-range nuclear interdiction might be too long-term to affect the immediate outcome. In this case, the possession of long-range forces designed to effect reprisals against the use by the enemy of his own long-range weapons may be considered a better posture for the Alliance to adopt.

The actual impact of a modernised LRTNF arsenal on the outcome of a nuclear battle in Europe is thus one obvious question that arises in connection with the modernisation debate. To the extent that improved long-range nuclear forces would enable the Alliance to achieve some battlefield advantage, such as halting an attack, or stabilising the Alliance's conventional defences, then their effect would be to strengthen that view of escalation which sees the process as holding out the possibility of intra-war deterrence. On the other hand, even if modernised LRTNF did not give any clear escalation advantage to NATO by immediately affecting the outcome of the fighting at the forward edge of the battle area (FEBA), the ability to threaten reprisals in the event of a pre-emptive, counterforce attack would reinforce the total spectrum of nuclear options making up the Alliance's deterrent posture. That is, escalation is seen as a means of linking theatre defence with the American strategic deterrent. In either case, at the very least, a capacity to undertake long-range theatre nuclear strikes can be considered as adding to the number of possible responses the Alliance could make in the event of conflict in Europe.

Arguments in favour of the deployment of modernised long-range theatre forces have tended to emphasise the latter aspect of the contribution that these forces could make to deterrence in Europe. However, in addition to increasing the variety of military options available to NATO, a number of additional arguments have been put forward which, taken together, constitute a strategic rationale for the deployment of modernised LRTNF. One argument, which has been made much of by official spokesmen, is simply that the increase in Soviet battlefield and long-range theatre nuclear capabilities should be matched in some way. The NATO arsenal does not include land-based missile forces in any way comparable to the Soviet SS-20 missile, and the forward based systems presently deployed on behalf of the Alliance are increasing in age and vulnerability. The modernisation of the Alliance's long-range theatre nuclear weapons would thus be a response to the improved Soviet theatre nuclear capabilities and the increased threat to Western Europe that they represent.[3]

It is also argued that the trend towards Soviet theatre nuclear superiority (which some believe has already been conceded), could undermine the stability achieved in intercontinental systems and highlight the 'gap' in the spectrum of possible NATO responses to aggression.[4] Presumably what is meant here, is that in conditions of strategic parity, theatre inferiority may undermine Alliance confidence in the level of security provided by American forces 'essentially equivalent' to those of the Soviet Union. When American forces were in some way measurably superior to those of the Soviet Union, it could be argued that the size and character of Soviet long-range theatre forces were less significant to Western security, since any use of LRTNF by the Soviet Union would carry the great risk of a response from superior American strategic forces. With the loss of strategic superiority, doubts arise as to the deterrent effectiveness of strategic weapons as a counter to Soviet Eurostrategic forces, and consequently there is an increase in the importance of the Alliance's own long-range theatre nuclear weapons. A 'gap' thus arises if there is no credible military response available to the Alliance at the long-range theatre nuclear level.[5]

Of course, the credibility of the American strategic commitment was questioned long before parity obtained in the central strategic balance. The movement of the United States in the late 1960s towards a strategy of assured destruction can be interpreted as a recognition of the limited 'usability' of strategic superiority in the defence of Western Europe. Indeed, it is possible to interpret the Alliance doctrine of flexible response, adopted in parallel with the shift in the American strategic posture, as a means of compensating for the reduced credibility of that commitment in a situation of increasing American vulnerability to Soviet reprisals. A vital function of the Alliance's theatre nuclear forces became that of linking the defensive posture in Europe with American strategic weapons by providing an alternative to strategic nuclear response in the event of a localised theatre attack.

Given the character of Soviet Eurostrategic forces at that time, and the level of destruction that their use would cause, American strategic weapons were still considered as a credible retaliatory threat to them. Thus the American implementation of the strategy of flexible response tended to stress the deployment of shorter range battlefield and tactical weapons in Europe. As a result of these doctrinal and postural changes, the strategic significance of the Alliance's theatre nuclear weapons increased; they became essential components of a strategy of escalation. This was accompanied by an easy assumption that NATO possessed theatre nuclear superiority, and that the presence of tactical nuclear weapons in the European theatre favoured NATO more than the Warsaw Pact. The Soviet theatre build-up has called this assumption into question, and this represents a fundamental challenge to declaratory Alliance strategy.

In a situation in which any Alliance theatre nuclear superiority has been eliminated, the effect of NATO deploying a modernised force of long-range nuclear weapons would lie not so much in the restoration of superiority in this area, as in the denial of a comparable advantage to the Soviet Union. In present circumstances, modernised LRTNF may be a means of ensuring parity at the long-range theatre nuclear level in the manner publicised by Helmut Schmidt in his 1977 Alastair Buchan memorial lecture.[6] The maintenance of parity in the European theatre is desirable not only as a direct counter to the equivalent systems of the Soviet Union, but also for their deterrent effect on the adversary's other options. In this view, LRTNF are essential to the 'seamless web' theory of deterrence in which all elements in the NATO strategic posture are mutually reinforcing in their deterrent effect. However, within the framework of the strategy of flexible response, there are two additional considerations that affect any judgement about the contribution modernised LRTNF make to the reinforcement of the Alliance's deterrent posture: first, whether the threat to initiate the use of nuclear weapons any longer has credibility and secondly, whether the Alliance has options, other than the use of strategic forces, that might deter the Soviet Union itself from initiating the use of nuclear weapons.

Strategic parity and the adverse trend in the theatre balance have further reduced the desirability and credibility of an Alliance posture premissed on the possibility of initiating the use of nuclear weapons. Among the objections to first-use is the argument that if the initial use of nuclear weapons were made in the face of conventional defeat, then the nuclear battle would almost certainly take place on allied territory. Use on anything other than the most restricted scale would risk enormous damage to the defenders' own territory. In any case, if the first-use of nuclear weapons failed to halt the enemy and cause him to reconsider the consequences of his attack, then the attacker can be expected to respond in kind. Indeed, a limited first-use on the part of the Alliance would seem to invite the Warsaw Pact to attempt the pre-emptive elimination of NATO's remaining theatre nuclear forces.

In order to meet these objections, and as an alternative to initial battlefield use, the suggestion has been made from time to time that NATO might resort to long-range interdiction. It can be argued that this would have considerable demonstration effect, and that it would have the advantage of not involving damage to allied territory. Ultimately, of course, a strategy of this kind would work only if the adversary were deterred through fear of further escalation. If the Soviet Union were to respond with similar long-range forces, then this would increase the risk of engaging strategic weapons. This risk, however slight, must always be a powerful deterrent to the Soviet Union's use of Eurostrategic weapons. None the less, does the commitment of strategic forces any longer have sufficient credibility to the allies themselves for them to use it as a means of escalation control after the Alliance had initiated the use of nuclear weapons?

In so far as the Alliance wishes to retain an option to initiate long-range interdiction, the ability to withhold LRTNF in a retaliatory role would seem to be necessary. In the past, the undoubted second-strike characteristics of American strategic weapons would have been held to perform this function. Now, in the light of strategic parity, it must be accepted that to some extent at least these forces have been neutralised. The possession by the Alliance of relatively invulnerable LRTNF could deter the Soviet Union from an escalatory response that might in turn place its own territory at risk. In any event, the Alliance's LRTNF serve to deny the Soviet Union the prospect of maintaining its own territory as a nuclear sanctuary in the event of large-scale conflict in Europe that did not invoke the employment of American strategic forces.

Not least among the arguments in favour of the Alliance acquiring a modernised long-range theatre nuclear arsenal, is that it will be more secure against pre-emption than existing NATO forces. Regardless of any demonstrative effect the initial use of such forces might have, and discounting the possible value of any improved counterforce capabilities, confidence in their ability to survive a disarming strike would appear to be a minimum requirement of a strategy of flexible response. For the possibility of flexible response to be sustained, it is important that the Alliance's theatre nuclear weapons do not present a tempting target for pre-emption. Just as so much attention has been given to the survivability of strategic forces, so a situation of strategic parity makes it desirable to apply more rigorously to theatre weapons those concepts concerning survivability and intra-war deterrence that were developed in connection with the strategic forces. For this, the Alliance should have confidence in its ability to withhold theatre strike forces as a continuing deterrent to the exploitation by the Soviet Union of its own theatre nuclear weapons. By so doing, the Alliance holds out the prospect of being able to exercise escalation control in the event of conflict.[7]

An important consideration arising here is whether the Alliance is any longer in a position to exercise any kind of escalation control, and whether LRTNF modernisation would make any difference in this respect. First of all, if the problem is seen as one of maintaining a credible and stable posture of extended deterrence, then the contradiction between strategic forces governed by a doctrine of assured destruction and a theatre doctrine of flexible response has to be resolved. It is one thing to use the concept of assured destruction as a guide to minimum force posture requirements when possessing a substantial superiority in strategic capability, but it takes on a different significance when these requirements are established within the framework of 'essential equivalence' or 'sufficiency'. When the criteria for assured destruction were being established for American strategic forces in the 1960s, the United States still possessed a 'surplus' capacity that could be allocated to the targeting needs of flexible response in Europe. Now, however, despite the continuing targeting commitment and the renewed emphasis on limited options in recent American posture statements, the United States no longer possesses the variety of strategic targeting options, for example, that enabled Robert McNamara, as Secretary of Defense in the early 1960s, to suggest the possibility of unilaterally imposing damage limiting and city avoidance strategies. Indeed, there is scepticism as to whether (particularly in the light of growing Minuteman vulnerability), the United States possesses forces capable of meeting the requirements of a full-fledged limited options posture.[8]

A capacity for assured destruction may give the Americans a high degree of confidence in their ability to deter a Soviet first-strike on themselves. But it is problematic whether the strategic forces of the United States, as presently constituted, could exercise escalation control in the event of a conflict in Europe that could not be dealt with by theatre forces. It is not so much that the American strategic commitment lacks credibility as a deterrent to a Soviet attack on European cities. This would be an extreme action that could well precipitate general war. Rather, it is a question of whether the threat to use strategic weapons against Soviet targets would serve to deter the Soviet Union from exploiting any tactical and territorial advantages that might have been secured at lower levels of violence once hostilities had broken out.

Presumably, in order to exercise escalation control, the Alliance should obtain some actual or potential military advantage from the process of escalation; however, given the limited counterforce capabilities of American strategic forces against the emerging Soviet Eurostrategic threat to Western Europe, it is difficult to see how the initiation of a limited counterforce attack would leave the United States in a superior military position *vis-à-vis* the Soviet Union. The problem in this case is that the present structure of strategic parity would seem to give the Soviet Union the advantage in any counterforce exchange with the United States. The exchange ratio would be adverse for the United States. If the United States were to initiate a selective counterforce strike, the present vulnerability of the Minuteman force, which is the main counterforce component of the American strategic triad, would invite the Soviet Union to attempt the pre-emption of the remaining Minuteman missiles. Quite apart from the horrifying casualties that would result from such an attack, the United States would find itself in a position in which the Soviet Union still possessed counterforce options, while the United States was restricted to a countervalue assured destruction capability. This force would continue to confer security on American cities, but would be unable to affect hostilities in Europe, or redress an unfavourable Eurostrategic balance.[9]

In this situation it can be argued that the strategy of flexible response is in effect restricted to the theatre level. Once the level of theatre nuclear operations had been reached, and if 'escalation control' had not been successfully achieved, then it is doubtful if the limited use of strategic weapons would justify the risks to the United States. This doubt is reinforced by the limited counterforce capabilities against the new Russian Eurostrategic systems of the Poseidon missiles assigned to SACEUR. This provides one argument for those who advocate that, even within the framework of SALT, the United States should acquire improved strategic counterforce capabilities. However, in the absence of these forces, the potential of modernised LRTNF is thus canvassed as a means of redressing the balance of mutual deterrent threats by closing the 'gap' in the spectrum of deterrence, and by ensuring that Soviet territory is vulnerable to attack by theatre forces.

There is a paradox here though; for in so far as there is a gap in the spectrum of deterrence, this arises from the degree of what has been termed 'strategic disconnection' that has occured as a result of the shift in the strategic balance.[10] On the other hand, a modernised, long-range theatre nuclear force of a size and character to offer a comprehensive second strike, counterforce capability against Soviet LRTNF would itself be decoupling in effect. What the Alliance would be seeking in this case would be a Eurostrategic balance based upon strategically autonomous theatre capabilities. (The question of how these forces would be controlled is a significant, but separate, issue.) Certainly, the scale of the proposed Pershing II and GLCM deployments is not sufficient to establish such a separate balance, quite apart from the fact that at present there is not the political possibility of securing agreement to do so. In other words, the deterrent effect of the proposed deployments continues to rest on their being linked to American strategic forces that are themselves increasingly limited in the range of threats that they can deter. To pursue the paradox further, modernised LRTNF are dependent on the very forces whose weaknesses they are supposed to alleviate by their own deployment.

One conclusion that can be drawn from this is that the deployment of modernised LRTNF cannot alter the fact that in the final analysis their deterrent effect is only as reliable as the American strategic guarantee. To the extent that these weapons remain American controlled this is undoubtedly so; but the reliability of that guarantee is not simply the result of a constantly reaffirmed act of will, nor of the fact that miscalculation about the consequences of conflict in Europe is likely to be disastrous for all concerned. Without placing too much burden on the concept of 'rationality', it can be recognised that the balance of interests as perceived by the responsible authorities in time of crisis will also crucially affect how that guarantee is interpreted. The possibility of using European-based LRTNF for tasks that might otherwise involve the use of strategic weapons, may well affect how the balance of interests and risks is seen by an American President and his advisors. Ironically, a degree of strategic disconnection is necessary if the American strategic guarantee is to have any credibility at all; thus the nice point to establish is whether any level of LRTNF deployment would have disutility for Western Europe security by weakening the American strategic umbrella over European cities.

This raises the question of whether the augmenting of the Alliance's LRTNF resources would weaken, rather than reinforce, the connection between theatre and strategic nuclear forces. However, this question cannot be settled by force characteristics alone. The strength and nature of the linkage will depend rather on the strategic doctrine governing the potential use of modernised LRTNF, and on the political-psychological responses of governments to their deployment. Thus a number of European commentators have treated the proposal to deploy long-range Pershings and GLCMs with suspicion. They have argued that instead of strengthening the ladder of escalation, the reinforcement of the Alliance's LRTNF capabilities would make it easier to confine any nuclear exchange to European territory. This falls into that class of arguments which suggest that the strategic trend has been towards the separation of the conditions of West European and North American security. No longer can the security of the North Atlantic area be considered indivisible, and therefore nothing should be done which might widen this division. The trouble with these arguments is that if the operational strategy of the Alliance (as distinct from the declaratory strategy) was to respond in the same way to any particular level of attack on any of the various allies, then the American strategic commitment would be unacceptably dangerous to the United States. No American president can commit himself in advance to particular responses to hypothetical contingencies; however, what can be done is to provide the wherewithal for a plausible range of responses to various kinds and levels of conflict in Europe. The possibility of a retaliatory strike against the Soviet Union with theatre-based weapons is one such option, but one which is presently represented by an aging and increasingly vulnerable arsenal of forward based systems.

The deployment of 108 Pershing II missiles and 464 GLCMs is far short of the numbers that would be required to match on a one-for-one basis the Soviet long-range forces directed against Western Europe. In the light of the strategic questions already discussed, the question thus arises as to the purposes that could be served by the new deployments.

If for no other reason than to avoid giving credence to the notion of a distinct Eurostrategic balance, allied spokesmen have made it clear that the new deployments are not intended to offset directly particular components of Warsaw Pact forces. Instead, the new missiles should be seen as a response to adverse developments in the Alliance's strategic environment, and their deployment as a means of strengthening the overall deterrent posture of NATO.[11] One means by which this can be done is by strengthening the Alliance militarily. The new missiles would replace at least some of the Alliance's increasingly vulnerable 'quick reaction alert' (QRA) aircraft, and release dual-capable aircraft to a conventional role. As well, both the Pershing II and the GLCM offer the much greater prospect of penetrating the very strong air defences of Eastern Europe and the Soviet Union. Against Soviet targets in particular, the main forward based systems currently available to SACEUR (the British V-Bombers and the American F-111s) are neither numerous enough, nor invulnerable enough, to be confidently withheld as a counter to Soviet LRTNF. Provided that they do in fact meet the requirements of 'invulnerability' and 'penetrability', the modernised weapons would be able to perform better the roles for which the current forward-based systems are presently deployed.

The Pershing II extended-range missile (Pershing II XR) is a development of the Pershing II missile tested during 1977 and 1978. The original Pershing II had a similar 400 nautical mile range to the Pershing

Ia, but with a terminally guided re-entry vehicle having a reported CEP of about 120 feet, which is a tenfold increase in accuracy over the earlier missile. This order of accuracy would give the missile a highly effective capability against hardened, high value targets, and would also allow for the employment of relatively low-yield warheads with consequently lower amounts of collateral damage.[12] The extended range version of the Pershing II currently under development has a range of about 1,000 nautical miles and, like its predecessors, is mounted on mobile launchers. Operational deployment could be expected as early as 1983-4.

The proposed ground-launched cruise missiles would also feature the great accuracy associated with terminal guidance, and since they would be mounted on specially designed vehicles, also offer the advantages of mobility. Given, however, that they would fly at subsonic speeds, the time they would take to reach their destinations reduces their effectiveness against time-urgent targets. Nevertheless, their accuracy makes them suitable for targeting against a wide variety of hardened and non-hardened military objectives, and as with the Pershing II, lower yield warheads than those employed with existing systems could be used.[13] The GLCM will have a range of 2,500 km.

The task of modernised LRTNF has been described as that of attacking high priority fixed or semi-fixed land based targets normally located well to the rear of the forward edge of the battle area. These targets would include IRBM/MRBM sites, naval, air and ground force bases, as well as headquarters and C^3 complexes, 'choke points' and bridges. Overall, the intention would be to weaken or deny the Warsaw Pact the ability to conduct sustained military operations.[14] Despite the revisions in targeting that the deployment of new LTRNF will undoubtedly bring about, it can be presumed that a good percentage of the force would be employed against targets outside the Soviet Union. Since, presumably, the targeting of theatre nuclear weapons against objectives in the Soviet Union would be integrated with that of strategic weapons assigned to similar objectives, it is difficult to estimate what proportion of modernised LRTNF would in fact directly threaten Soviet targets. What can be deduced is that, potentially, the new weapons represent an increased threat to Soviet territory, and that in all probability, LRTNF would either take over some targets presently assigned to strategic weapons, or generally increase the coverage of military targets in the Soviet Union that are seen as being of direct interest to the European allies. It is also possible to conclude that the deployment of the Pershing II and GLCM will increase the possibility of employing limited nuclear options in the European theatre.

None the less, the performance characteristics and size of the planned deployments indicate that no dramatic change in the military role of the LRTNF available to NATO is planned. Essentially, they would continue to contribute to the three kinds of military tasks that are presently assigned to theatre nuclear forces. These have been described as limited nuclear options designed to permit the selective destruction of fixed military and industrial targets; regional nuclear options, which would be restricted to a specific area where targets were located, for example the leading units in an attacking force; and theatre-wide nuclear options, which would be directed at aircraft and missile bases, first and follow-on echelons of any enemy attack and general logistical interdiction.[15] However, the scale of the planned modernisation would seem to be insufficient to enable military planning to proceed on the basis of the greatly extended use of nuclear weapons against second and third echelon forces, and it does not seem likely that the new deployments would meet the military requirements of the full nuclear war-fighting posture for NATO that some proponents of theatre nuclear force modernisation have advocated.

Still, the new weapons do hold out the prospect of more discriminate use than does the existing arsenal. The possibility of discriminate use increases the chances of maintaining some restraints once nuclear escalation had occurred. Even if the new weapons do not restore 'escalation control' to Alliance forces, they may nevertheless strengthen the credibility of the escalatory threat underlying the Alliance's deterrent posture by providing for more effective follow-on options once the Alliance had resorted to some form of limited nuclear use.[16] There is no way the Alliance can escape from the possibility of a devastating Soviet counter-response, but it can place the onus of further escalation, with all the risks that this entails, on the Soviet Union. Again, this would not necessarily increase the credibility of NATO first-use (though it would not necessarily reduce it either), but it would provide the Alliance with increased options in the event of the Warsaw Pact initiating a nuclear conflict. At the very least, modernised LRTNF would provide for the more effective implementation of existing military doctrine, which is not, moreover, without its political and psychological value as long as the Alliance is committed to a strategy of flexible response.

Notes

The author was assisted in the preparation of this chapter by the award of a NATO Fellowship in 1979. The assistance of the Research Board of the University of Manitoba is also acknowledged.

1. *Communique*, Special Meeting of NATO Foreign and Defence Ministers, Brussels, 12 December 1979. It was subsequently revealed that the planned distribution of GLCMs would be as follows: the United Kingdom would provide bases for 160 missiles, Italy 112, Germany 69 and Belgium and the Netherlands would each base 48 missiles.

2. A variety of estimates have been made of the balance of theatre nuclear forces. Two detailed examples can be found in Robert Metzger and Robert Doty, 'Arms Control Enters the Gray Area', *International Security*, vol. 3, no. 3 (Winter 1978/9), pp. 17-42; also, *The Military Balance 1979-80*, (IISS, London, 1979), pp. 114-19. The latter estimate uses a definition of LRTNF that is broader than this author would accept. The essential political and strategic significance of LRTNF is that they be 'Eurostrategic' in capability. That is, on the Western side, that they be capable of striking targets in the Soviet Union from bases in the European theatre. The IISS estimates include NATO systems that are not capable of being deployed operationally in such a role. However, the 1980-1 edition of *The Military Balance* makes a clearer distinction between LRTNF and theatre nuclear forces generally, defining LRTNF as forces with a range over 1,000 km.

3. *Communique*, Special Meeting of NATO Foreign and Defence Ministers.

4. Ibid.

5. Stephen R. Hanmer Jr., 'NATO's Long-Range Theatre Nuclear Forces', *NATO Review*, vol. 28, no. 1 (February 1980), p. 1.

6. Reprinted in *Survival*, vol. XX, no. 1 (January-February 1978), pp. 2-10.

7. *Report of the Secretary of Defense*, Harold Brown to the Congress on the FY 1980 Budget, FY 1981 Authorizaticn Request and FY 1980-84 Defense Programs, 25 January 1979, p. 84.

8. For example, Colin S. Gray, 'Nuclear Strategy: A Case for Victory'. *International Security*, vol. 4, no. 1 (Summer 1979), p. 64. The development of the so-called 'countervailing strategy' has not altered this basic state of affairs. See, *Report of the Secretary of Defense*, Harold Brown to the Congress in the FY 1981 Budget, FY 1982 Authorization Request and FY 1981-1985 Defense Programs, 29 January 1980, pp. 65-6.

9. Of course, the results of any counterforce exchange between the United States and the Soviet Union are subject to many uncertainties; not least because of the inevitably arbitrary character of many of the assumptions on which calculations are based. For arguments suggesting that the Soviet Union would emerge stronger than the United States after a counterforce exchange see: Edward N. Luttwak, *Washington Papers*, IV, (Sage, Beverley Hills, California, 1976), pp. 61-2; Paul H. Nitze, 'Deterring our Deterrent', *Foreign Policy*, no. 25 (Winter 1976/7, pp. 195-210. A good discussion of the factors affecting such evaluations can be found in *Counterforce Issues for the US Strategic Forces*, Background paper, Congressional Budget Office (US GPO, Washington, D.C., January 1978).

10. Pierre M. Gallois, 'NATO's Obsolete Concepts', *Conflict Studies*, no. 96 (Institute for the Study of Conflict, London, June 1978), p. 6.

11. See, for example, 'Modernization and Arms Control for Long-Range Theatre Nuclear Forces', United States International Communications Agency (USICA), *Policy Backgrounder*, 8 November 1979, p. 6.

12. *Aviation Week & Space Technology*, 15 May 1978, pp. 19-20.

13. *Fiscal Year 1980 Arms Control Impact Statement* (US GPO, Washington, 1979), pp. 136-7.

14. Ibid., p. 133.
15. Brown, *Report of the Secretary of Defense, p. 85.*
16. Peter Stratmann and Rene Hermann, 'Limited Options, Escalation and the Central Region', in J.J. Holst and Uwe Nerlich (eds.), *Beyond Nuclear Deterrence* (Crane, Russak, New York, 1977), p. 253.

9 WHO IS DECOUPLING FROM WHOM? *OR* THIS TIME, THE WOLF IS HERE

Pierre Hassner

'Plus c'est la même chose, plus ça change.' To write on current events, and to do that by way of chapters and articles which appear about a year after having first been drafted, and which have to be revised in the meantime, provides one with a lesson both in humility and in stubbornness. In the spring of 1980, when addressing the conference which gave birth to this book, I started by referring to another conference which had taken place in the autumn of 1979, noting that the divergent Western reactions to the invasion of Afghanistan had revealed that the crisis in the Atlantic Alliance was deeper than I had thought seven months before. Another eleven months later, reactions to new crises – the Iran-Iraq war, and the dangers of a Soviet invasion of Poland – seem much less divergent, and more in tune with my balanced appraisal of 1979 than with my alarmist tone of April 1980. Yet there are enough grounds, inside and outside Europe, not to exclude the possibility that the calm which obtains at the time of this writing may only be a lull, as everybody holds his breath with Soviet troops poised on the borders of Poland and with a new Administration in Washington; the crisis over Afghanistan may yet turn out to be only a foretaste of a storm which continues to gather.

Can one go a little further than this sucession of immediate impressions and long-term question marks in order to ascertain the interplay between, on the one hand, structural problems associated with the nature of extended deterrence and arms control, and with the political interests of the two sides of the Atlantic, and, on the other hand, the accelerating crises in East-West relations and the varying Western reactions to them?

In Brussels, in September 1979, this writer's paper about 'Intra-alliance Diversities and Challenges' started with two slogans: 'Die Laze war noch nie so ernst' (The situation was never so serious) and 'Fluctuat nec mergitur' (It rocks but does not sink), and it remarked that 'both Adenauer's famous and annually repeated statement of alarm and the reassuring motto of the City of Paris have been equally characteristic of the discussion around the Atlantic Alliance ever since its inception'.[1] It noted that throughout this history three types of crisis had coexisted (crises over American protection or European independence, the discrepancy between domestic evolutions and alliance structures, and conflicts of interest outside NATO territory), and that after the worry about the Left and Eurocommunism had subsided, the main bone of contention was likely to be, as at the time of the Suez crisis and the Yom Kippur war, related to diverging responses to external challenges, particularly in the Middle East. It ended with the contention that while such conflicts could ruin the Alliance, a formal extension of the latter beyond its territorial and military core risked endangering a consensus which *did* exist over resistance to a direct Soviet military threat in Europe, while indirect challenges in the Third World brought out national differences between countries and ideological ones within them. I concluded: 'A little politics gets us away either from NATO or from its strictly defined defensive agenda. More politics brings us back to it.'

Looking back at this conclusion in April 1980, in the light of divergent European and American reactions to Iran and Afghanistan, I found it rather bland and complacent. Conversely, Irving Kristol's paper at the same conference, presented under the provocative title '*Does NATO exist?*' then seemed to me – in the spring of 1980 – much more attuned to the new realities than one had thought at first.[2]

Kristol's basic argument is that America is entering a new age of nationalism and global interventionism, while Europe is tired, social-democratic and inward-looking. Hence the latter will lose its priority for the former. Barring the unlikely hypothesis that they acquire the resolve and the unity to become a Superpower themselves, the Europeans have no choice but either to follow the United States in its renewed global role or become a matrix of South Koreas, i.e. of protected states anxiously worrying about the maintenance or withdrawal of American presence or protection. But the myth of an alliance based on common objectives and mutual trust will not survive the parting of the ways in the rest of the world; nor will America's willingness to run the greatest risks and commit its best forces on the continent where the action is *not*, for the sake of allies who are not willing to run any risk in supporting her struggle because of their own narrow interests.

It seemed, in September 1979, and still does, to some extent, that this picture, while containing a grain of truth, was vastly overdrawn. The United States had not entirely forgotten the lessons of Vietnam and its new interventionism remained, up to that point, largely rhetorical. In Europe, the crisis of social democracy, the move to the right, the disillusionment with detente and the preoccupation with defence, while by no means universal, were at least observable secular trends in several key countries.

Nothing which has happened since has invalidated these objections. Indeed, some of them have been confirmed by the weakness of some of the reactions of the US government (about aid to Afghan resistance for example), the attitudes of candidate Reagan (about the grain embargo, the Olympics and the draft), and the hardening of some European governments, such as France. But Iran, Afghanistan and El Salvador have brought to the fore such a degree of differences in attitude towards East and South, such depths of distrust and misunderstanding amongst each other that the more technical disagreements about deterrence and arms control, about SALT and TNF modernisation, were looking abstract and almost irrelevant by the spring of 1980. The decisive level of analysis did indeed seem to be the one emphasised by Kristol, i.e. the discrepancy in psychological attitudes between the United States and its European allies.

On the other hand, it is remarkable that the 1980 Presidential election, which, in the long run, should both dramatise and deepen the gap in attitudes between the United States and most of its European allies, particularly the most important one, the Federal Republic of Germany, has been greeted in Europe with something of a sense of relief, if not with outright satisfaction, as among French Gaullists and British Conservatives. Even before that, Franco-American naval co-operation in the Persian Gulf, in view of the Iran-Iraq war, was judged by both sides as satisfactory. Moreover the general European inclination to treat the invasion of Afghanistan as a distant, regional problem was much less in evidence in this latter case, even though the Federal Republic was not any more inclined to get directly involved. And on Poland, there was an almost perfect consensus in the autumn of 1980 between President Carter and Chancellor Schmidt on avoiding any provocative actions while assisting the democratisation of Polish society through economic means.

One may indeed think that in both cases – Poland and the Gulf War – consensus was more easily reached than after Aghanistan because everybody agreed that there was not much which could usefully be done. A Soviet invasion or, more particularly, an indirect or ambiguous intervention may well bring a new polarisation. Yet in December, the hardening of the American position through public warnings directed towards the Soviet Union was matched by a similar attitude coming from the Luxembourg meeting of the Nine. Again, the West Germans are one step behind, and their reluctance to risk losing the benefits of detente in terms of human contacts, economic benefits and security for Berlin, whatever happens in Poland, leaves entirely open the question of what their actual reaction would be.

One is left, then, more than ever, with a doubt about the roots of transatlantic differences, and, indeed, about whether the contours of these differences are even transatlantic in their essence. To some, the differences are above all differences in rational *assessments*, whether about the reasons for the Soviet invasion of Afghanistan, or about the likelihood of an invasion of Poland, or about the dangers of nuclear war. Indeed, while various countries are by no means unanimous on these various issues, the prevalent assessment of European Governments, on the invasion of Afghanistan for instance, began by considering it as a regional problem, and has since shifted to denouncing the danger of a Third World War through lack of communication between the Superpowers, whereas the prevalent American perception has focused on the danger of Soviet expansionism and shifts in the global and regional balance of power.

At the other extreme, one can see the root of transatlantic divergences less in differences of perception than in differences of *interest*. If Europeans are more sensitive to the dangers stemming from a closure of the Straits of Hormuz than from the invasion of Afghanistan, it is because their interests are more directly threatened. If they are more moved by the dangers to Poland, yet less eager to call an end to detente, it is, again, because their interests, human and economic, are more directly threatened. If the world has become a global theatre of conflict in which the American Superpower is necessarily involved, a regional structure of peace has, on the other hand, developed in Europe, and it involves European, particularly German, interests more than American ones. Similarly, the differences over deterrence have often been seen by Americans as stemming from differences in doctrine, or from an inadequate European understanding of strategic realities, whereas many Europeans have stressed the objective differences of interests between the United States and Europe concerning, for instance, the use of nuclear weapons in a limited war.

This brings us to the third interpretation, which lays primary emphasis on differences in *strategy*. According to a minority of Americans (like Robert Legvold),[3] Helmut Schmidt or Valery Giscard d'Estaing do not perceive the character of the Soviet threat or the meaning of the invasion of Afghanistan any differently than the United States; nor do they have any less interest in stopping Soviet expansionism, just as the continuation of detente is equally in the American interest. Rather, they disagree on the way to achieve these aims, with the Europeans favouring a more differentiated response which, instead of attempting to punish the Soviet Union across the board, would combine a tougher local resistance and restoration of the regional equilibrium in the Gulf area with a continuation of detente in areas, such as trade and arms control, where it is in the mutual interest.

While there undoubtedly is some truth in this view, I would personally combine it with an interpretation less charitable to the Europeans. To a great extent, the evidence, particularly as it appears in public opinion polls, shows that the image of the Soviet Union and the perception of its actions are no more favourable or reassuring in Europe than in the United States;[4] but the reaction to them and, in particular, the willingness to take risks, especially that of violent confrontation, in resisting them, vary dramatically. This is less a matter of rational strategy, however, than of moral and psychological attitudes, a perception of one's own situation more than of Soviet intentions. Very often an apparent optimism about the latter hides a deeper pessimism about one's self. Publicly or subconsciously, one is committed to the more reassuring interpretation of Soviet policies precisely because deep down, one feels one cannot in fact afford to act on the implications of the more worrying one.

Beyond the rational realm of assessments, interests and strategies, one finds, then, the more intangible and elusive – but even more essential – realm of *attitudes*, based on different or differently perceived *situations*. On a rational level, interests are more common than conflicting as between the two sides of the Atlantic, and some kind of consensus on the nature of the common threat and the best strategy to meet it should not be impossible to reach. But these very perceptions and strategies cannot help but be differently interpreted on the basis of deep-seated differences in attitudes, themselves based on differing historical experiences, leading to differing shades in each side's feelings of security or vulnerability.

This would help explain why, whereas detente has generally tended to encourage independent policies, and East-West crises have tended to reunify the Western Alliance, in 1980 the crises of Iran and Afghanistan tended to split it more than Gaullism or Ostpolitik ever did, and why, in the years ahead, the clashes over deterrence, American protection, East-West relations, or Soviet expansion may produce a continuing and deepening crisis within the Western Alliance related to the very meaning of detente and of arms control, and their demise or revival.

It is tempting, indeed, to interpret the present crises in the light of an instant philosophy of history, by distinguishing their effects upon three types of cycles. A well-known, short-term cycle in East-West relations is that of confrontation and detente. Usually a Soviet move is followed by a period of outrage and of coldness, itself followed by a resumption of detente a few months later, after a peace offensive. This is what happened after the Cuban missile crisis and the invasion of Czechoslovakia. But this time the pattern may be modified by an overlap with two different types of cycles.

One is the middle-range cycle of American foreign policy: after fifteen years of retreat, if one follows Samuel Huntington's calculations, the United States may be entering a new phase of activism. The events in Iran and Afghanistan may have given the decisive impetus to a sea-change which was already in the making for the last four or five years, thereby prompting American decisions which will have their full effect only in the second half of the decade but which may make present confrontations more likely (albeit in unfavourable circumstances) and a return to the old style of detente more difficult, as with the nascent crisis over El Salvador.

Conversely, in the case of Europe and Japan, there may be a secular trend, at work since 1945, towards becoming 'civilian powers without power', and against running any risks of war through active confrontation. In one case, the reaction to the crises may be harder than usual and exclude the usual return to detente; in the other the attachment to detente may have become unconditional, and exclude even the usual show of firmness and unity – however superficial and short-lived – during a crisis.

The immediate effect may well be that even if the Soviet Union were to withdraw from Afghanistan and dismantle all the SS-20s, the American turn towards a harder, more nationalist and more assertive foreign policy would continue, and that even if the USSR were to invade Poland there would be many important European, particularly German, voices which would claim that this is all the more reason to develop still more detente and trade, and not to fall into the trap of a return to the cold war.

The American illusion, shared, apparently, by many in the Reagan Administration, is that one can return to the cold war of the 1950s. The European illusion, shared, in particular, by many in social democratic governments, is that one can stick to the detente of the 1970s. Again, governments and, indeed, majorities may well hold more balanced and realistic views. But on both sides of the Atlantic there exist 'moral majorities' (which are, of course, but vocal and more or less influential minorities) who see politics in absolute terms, and who, in the American case, identify morality with the supremacy of the United States and the destruction of Communism, and which, in Protestant Europe, identify morality with peace at any cost, and immorality with nuclear weapons serving even in a deterrent function.

As for the more pragmatic general public, recent polls indicate a wide gap in attitudes towards the risks of war, with the Americans much more willing to resist agression militarily, even in Europe, than Europeans themselves.[5] This striking difference in attitudes is more likely to be modified in the direction of a lesser willingness of the Americans to run risks for Europe in a situation of nuclear parity, and of European passivity and transatlantic recrimination, than in the direction of greater solidarity. Everything seems to be ready for what may be called the *dialectics of decoupling*. What, in 1979, could appear to be somewhat academic discussion and speculation about the decline of extended deterrence, i.e. of America's protection of Europe, and about the desirability of extending the Alliance beyond the geographical borders prescribed by the Treaty, has become a subject of passionate controversy. Furthermore, at least in their handling of Iran and Afghanistan, each side has done its best to justify the worst apprehensions of the other.

For over twenty years, many Europeans had worried that, if there were to be a war, the United States and the Soviet Union would choose to wage it on European territory without using nuclear weapons or, at least, without using them on each other's territory. This was, of course, the main rationale for the French nuclear force. The protocol of SALT II had seemed to confirm this fear of 'sanctuarisation', and Henry Kissinger's speech in Brussels was taken as a confession of its truth. Conversely, in the same speech, Kissinger remarked that 'the secret dream of every European was, of course, to avoid a nuclear war, but, secondly, if there were to be a nuclear war, to have it conducted over their heads by the strategic forces of the United States and the Soviet Union'.[6]

Taken by itself, the TNF decision of 12 December 1979, to base, in Western Europe, American long-range missiles capable of reaching Soviet territory, should have gone a long way towards allaying this mutual fear of strategic decoupling. But as soon as the Afghanistan crisis broke M. Poniatowski, one of President Giscard d'Estaing's closest associates, rushed to confirm Henry Kissinger's suspicions by stating that the two Superpowers were bent on mutual suicide, and that Europe should get rid of NATO and build an independent nuclear force in order to stay out of their conflict. More importantly, the management of the crisis by the American Administration conveyed, rightly or wrongly, an image of alternating between weakness and recklessness, and the European governments took it as an alibi for their own passivity and equivocation. Hence whatever one may think of the reality or the dangers of *strategic decoupling* in the dimension of deterrence and defence, the temptation and reality of *political decoupling* in the dimension of diplomatic and economic relations with the Soviet bloc and with Iran are there for all to see. The same states, in particular the Germans, who fear American strategic decoupling and would like deterrence to be indivisible, hope for a political decoupling of intra-European and particularly intra-German relations from East-West ones, and would like detente to be divisible. Hence the crises of nuclear deterrence and of arms control which antedate the events of 1980 acquired a new gravity for external but no less deep reasons, just at the time when, on their own merits, a certain consensus, as exemplified by the TNF package deal of 12 December 1979, seemed to make it possible to live with them, if not solve them.

Of course, if somehow, as the present lull would authorise one to hope, the Alliance manages to live out the resentments and humiliations produced by everybody's behaviour in the South-West Asian confrontation and by divergent domestic evolutions, the objective factors which were beginning to produce a certain convergence in matters of deterrence and arms control, might continue to operate. Precisely in a world where defence issues are assuming a new reality, nobody can believe any longer either in pure nuclear deterrence or in an irreversible detente. According to the polls noted above, German and American public opinion have both tended to value detente less and NATO, particularly America's presence in Europe, more. The French, without abandoning their basic doctrinal and political positions, have at least entered the intellectual universe both of graduated deterrence and of arms control, with the discussions of their plans for the modernisation of their nuclear force on the one hand, and a conference on disarmament in Europe on the other. Everybody knows that a European nuclear defence is impossible today, and everybody feels, however confusedly, that it is indispensable in the long run.

The very essence of the subject, then, lies in the complex interplay between the specific logic and dynamics of deterrence and arms control, and those of political processes within the countries of the Alliance, as well as with each other and with the rest of the world. The crises of nuclear deterrence and arms control interact not only with those of East-West relations – exemplified and exacerbated by Afghanistan – but also with those of North-South relations, currently focused on oil, while previously preoccupied with the American hostages in Iran.

Here again, perhaps the main problem lies in the interaction of the two. The hostage crisis was not primarily an East-West incident; it was the first manifestation of a new type of crisis where both force and negotiation are made almost impossible by the elusive character of the opponent, and by the absence of any common, not only moral or legal, but even pragmatic ground. Anarchy and unpredictability appear as the main danger – a 'Khomeinisation' of the world rather than its communisation.

But, on the other hand, the meaning of Afghanistan is that it brings to the fore the presence of a very traditional danger – that of Soviet military expansion – threatening the West, its military position and its oil resources. But the incalculable element introduced by this traditionalist resurgence also threatens the Soviet Union, and Soviet expansion also threatens its Islamic neighbours. On the other hand, Western weakness towards Khomeini's challenge may have encouraged Soviet adventurism, just as Western military action risks throwing Iran and Islam into Soviet arms, or at least blurring the salutory regional effect of the invasion of Afghanistan.

To understand these complex interactions and, more importantly, their effect on relations within the Western Alliance, it may be advisable, rather than engaging in a substantive analysis of these crises, briefly to examine their roots and consequences at the level of certain broad foundations upon which both the solidity and fragility of NATO and detente are based. These relate to perceptions of *continuity* and *discontinuity*, *symmetry* and *asymmetry*, *vulnerability* and *invulnerability*, *predictability* and *unpredictability*.

Continuity and discontinuity

Discussions about extended deterrence, about the extension of NATO and the involvement or solidarity of allied states in non-treaty contingencies, about the Soviet sphere of influence and the projection of power, are full of implicit perceptions and assumptions of continuity and discontinuity. This can be taken in several senses: geographic (as with the distinction between a continental alliance or empire, and one extending across oceans); strategic (as between steps on the escalation ladder or criteria for the limitation of war, where thresholds can be technological as well as territorial); and political (as with the most recent and dynamic form of the Brezhnev Doctrine, where ideological or political continuity, through past association, the granting of the 'socialist' label, or a friendship treaty, seem to make countries eligible for fraternal help, and to make the conquests of socialism irreversible).

Of course these three meanings are linked, and one of the functions of treaties and, even more, of physical tokens of commitment like human hostages and the basing of nuclear weapons, is to compensate for geographical discontinuity by creating a strategic and political continuity. This is one of the roots of the mixture of objective conflicts and subjective misunderstandings which have plagued controversies over NATO deterrence and defence. Some – particularly the French – tend to see these differences as essentially clashes of interest; some – particularly the Americans – prefer to blame differences of doctrine or, rather, differences in degrees of comprehension of the 'true doctrine'. As indicated above, I prefer to see them as differences of perception rooted in differences of situation.

Americans tend to emphasise continuities as derivative of the idea that modern technology and communications have rendered traditional notions of distance meaningless.[7] The French tend to emphasise the discontinuity between the American territory and the European one (the United States being, to use Colin Gray's expression, a power in Europe, whereas Russia is a European power, and the American presence in Europe being a phase, admittedly a long one, but still a phase, of US foreign policy).[8] However, while the discontinuity between the sanctuary of American territory and that of the European allies seems quite clearcut in the French view, it is less clear whether it is due to geography (which would make it asymmetrical with the Soviet Union's), to Superpower status or to the ontological distinction between a nation's territory and that of its neighbours or allies. Hence an ambiguity in their own case: more than any country, they base their defence policy on the 'sanctuarization' of their national territory, yet occasionally (and increasingly) they stress its geopolitical continuity with the rest of continental Europe, particularly Germany.

The Germans, while worried about the discontinuities between the European theatre and the Superpower balance, waver between aspiring to a Eurostrategic balance (but still based on America's presence), and fearing its decoupling implications; on balance, they tend to emphasise the continuity of deterrence and hence of the escalation ladder. This is also the basis of NATO deterrence as seen from the American side.

However, it is here that an inevitably ambiguous combination of continuity and discontinuity appears. Both Americans and Europeans have an interest in deterring a war in Europe, whether nuclear or conventional, and in preventing it from becoming totally suicidal if it does happen. But for the Europeans the difference between the two terms is less than for the Americans; the former have an interest in accepting a bigger risk of total nuclear war and the latter a bigger one of limited European war. While, from the point of view of deterrence, both have an interest in proclaiming the continuity of the escalation ladder, if deterrence were to fail the Americans would have an interest in seeking to maintain the discontinuity in order to limit escalation. (So, in fact, might some Europeans when faced with the actual decision.) The point is that they couldn't be sure to succeed. The very presence of their troops and nuclear weapons creates, from the point of view of an eventual aggressor, an element of entanglement or continuity even though the defender may want to disengage at the hour of truth.

Hence the essential ambiguity of every threshold which, as Albert Wohlstetter has written about tactical nuclear weapons, can be interpreted either as a firebreak or as a bridge to a higher level.[9] In each case, there is a risk (or a chance) of decoupling because the United States will tend to limit the conflict to the lower level or to disengage from risks it does not control. In each case there is a risk (or a chance) of escalation, hence there is a recoupling from the point of view of deterrence, since every time a taboo (that of war as such, or of atomic weapons, or of Soviet territory) is broken, there is an increased risk of going to the next highest level.

While this is likely to remain valid, the evolution of technology, of the East-West balance and of political relations within the West can make a significant difference. It is just as absurd to think (as many misguided interpreters of Kissinger's speech in Brussels do) that in spite of the presence of hundreds of thousands of American troops, of thousands of tactical nuclear weapons of which, eventually, hundreds may reach the Soviet territory, American extended deterrence over Europe has vanished, as to think (as some Americans like McGeorge Bundy still claim[10] and some Europeans still claim to believe) that because the ultimate hard core of American extended deterrence (a second-strike counter-city capability against the Soviet Union, plus American hostages in Europe) is still there, nothing has changed in spite of the shift in the balance of vulnerabilities both at the intercontinental and at the continental level. What this shift means is that the United States has lost the control of escalation to the extent it used to have it, and that, to the extent it still has some choice, it will be even more reluctant to escalate. Since, on the other hand, the confidence of Europeans in American decisions has decreased for other reasons, the risks of strategic and of political decoupling feed upon each other. The question is not whether the United States will defend Europe but *how*, and *who* will decide upon strategic choices. Similarly the question about SALT is not whether it is conducted against European interests or ignores them (although SALT II came pretty close to doing so) but *how* they take European interests into account.

In both cases the discontinuity between the two Superpowers and their allies has tended to grow. Curiously this has been more evident in diplomatic than in military practice: the consideration of the Backfire only to the extent that it can reach the United States, and the exclusion of the SS-20 (which some believe has been deployed precisely as an effect of SALT, thereby resurrecting an old Soviet pattern of giving medium-range missiles priority over intercontinental ones and taking Western Europe as a hostage) imply a *de facto* acceptance of the Soviet definition of 'strategic' (what can reach the territory of one of the Superpowers). Added to the 600 km limit (which means a separation between the territory of Eastern Europe and of the Soviet Union) on GLCMs and SLCMs, this is the first confirmation of French theories of Superpower sanctuarisation of their respective territories. Conversely, the reluctance of certain Europeans to accept American nuclear weapons on their soil as a remedy to a Eurostrategic imbalance which they were the first to denounce suggests an attempt at political decoupling from the military East-West balance in order to pursue a separate detente with the Soviet Union or with Eastern Europe.

This is why the decisions about long-range theatre nuclear forces are so important. They could, of course, have had a 'decoupling' effect if they had aimed at an autonomous strategic balance and/or if they had coincided with an American acceptance of inferiority and an 'assured destruction only' posture at the intercontinental level. But since the numbers involved certainly run no danger – to say the least – of making the American intercontinental role superfluous, and to the extent that a parallel effort *is* being made at that level, they do have the essential effect of re-establishing a greater degree of continuity between the four territories: those of the two Superpowers and those of the two halves of Europe. Those who claim they involve no change as compared to missiles based in the United States or, conversely, as compared to existing tactical nuclear weapons, or that they would do a better job if put at sea, neglect their essential function, i.e. the hostage one. It is precisely the acceptance by the United States and by Germany of a greater risk of escalation to the intercontinental level which constitutes the 'recoupling' factor. It is precisely because these missiles are neither totally vulnerable nor totally invulnerable that a Soviet aggressor would have to try to eliminate them before launching an attack on Western Europe and would, hence, run the risk of inviting an American response on his own territory, and be faced with the dilemma of retaliating on American territory – or accepting a decisive asymmetry against itself.

Symmetry and asymmetry

The problem of continuity leads to that of symmetry. We shall abandon its other aspects (although the problem of the fixed or moving character of spheres of protection, influence and control is a crucial one today) to focus on the problems which the difference in the situation of Western Europe as compared to Eastern Europe in terms of dependence on and continuity with their respective Superpowers creates for the former, and for East-West relations and arms control in general.

But first it must be noted that already at the bilateral US-Soviet level, the crisis of arms control and that of detente are primarily tied to the discovery, on the Western side, of essential asymmetries which had been ignored. Of course the advocates of detente were always careful not to identify it with convergence, and arms controllers, even when inspired by the theory of games, wanted precisely, like Thomas Schelling, to abandon the symmetry postulates of the latter. However they did assume at least a partial symmetry in strategic postures and doctrines. Today, detente is crumbling under the Western realisation of the asymmetry between its own notion of the *status quo* and the Eastern conception of the 'dynamic *status quo*' (which President Kennedy had already translated in the formulation: 'What's mine is mine, what's yours is negotiable'). Arms control is crumbling because of the asymmetry in the conceptions of deterrence and stability of the two Superpowers which has made SALT as a mutually educating process a failure,[11] and because the asymmetry of their strategic postures makes it next to impossible to arrive at mutually satisfactory agreements, at least of the all-encompassing ambitions and proportions of SALT I and II.

But it is the particular situation of Western Europe which makes any conceivable agreement in SALT III necessarily lopsided. As Francois de

Rose has put it:

> If SALT III aims at overall parity in both medium and long-range weapons, the Soviet Union will be entitled to say that she is at a disadvantage, since all American weapons in that total could reach her territory, while only some of the Soviet weapons could hit the United States. If, on the other hand, the West agreed to the Soviet definition of strategic weapons – any weapons that could hit the territory of the Superpowers – Soviet missiles aimed at Europe would be excluded, while those that the United States might deploy in Europe would be included and thus subject to numerical limits. In that case, the US would have to accept the position of inferiority in central systems, giving up the policy which made SALT II possible. It is thus difficult to imagine negotiating aims that could at the same time be reasonable to the Soviet Union and militarily acceptable to the West.[12]

The ambiguity introduced in Superpower calculations by the existence of Western Europe produces the same irreconcilable differences in points of view in the current polemics about the TNF modernisation. The Soviets claim that an imbalance is produced because their warning time against an American attack launched from Europe is reduced to six minutes while that of the United States against an attack by them remains thirty-two minutes. If one left aside the absurd character of a Europe-launched American surprise attack (if only because of the numbers involved) and the existence of sea-launched missiles, the claim could have a certain plausibility in a purely bilateral context. After all, if the United States reacted with a world crisis against Soviet missiles aimed at its territory from Cuba, why should the Soviet Union not react the same way against American missiles aimed at its own territory from Europe? From a European point of view, however, this reasoning is of course completely unacceptable. The Europeans never have had more warning time against the Soviets than the Soviets had against them. If East Europeans have to declare that Pershing I missiles aimed at them are preferable to Pershing II missiles aimed at the Soviet Union, this is a measure of their lack of independence. But for West Europeans the idea that a middle-range missile is peaceful or tactical if it goes from Russia to Germany but warlike or strategic if it goes from Germany to Russia has little to recommend itself.

Of course, the Soviet bloc might apply the same reasoning to Cuba and, in a non-SALT context, respond to the NATO decision by putting

SS-20s there. What makes such a move – however far-fetched – interesting to consider at the present time, however, is another symmetry, concerning the events in Southern and Central Asia. Perhaps the most dangerous nightmare prompted by the Afghanistan and Iran crises is that of a Cuban missile crisis in reverse – with the United States engaging the Soviet Union at a time and place where the latter enjoys conventional superiority and where closeness to its borders creates an asymmetry in the balance of interests. In that case, to avoid or compensate for a humiliating retreat, the United States may, again, make an asymmetrical move and blockade or invade Cuba, as has already been suggested by such different people as George Kennan and Ronald Reagan. To which, in turn, the Soviets might retaliate on Berlin (as was feared during the original missile crisis) or negotiate the withdrawal of American missiles from Western Europe (as was perhaps tacitly done, then, with the Thors and Jupiters in Italy and Turkey).

Of course world politics is not played like a planetary chess-game of this kind, but this scenario does have the virtue of underscoring the symmetries between Western Europe, Cuba and, say, Pakistan, as exposed allies of one Superpower located in the vicinity of the other, and hence vulnerable to punishment by proxy aimed at their respective Big Brother.

Comparative vulnerability and generalised unpredictability

The word 'vulnerable' carries the somewhat intellectualised perceptions of continuity and discontinuity, symmetry and asymmetry back from the abstract world of nuclear strategy into the real world of the present crises. The analogy between the nuclear, the economic and the hostage situation is best expressed by the notion of *comparative vulnerability*. The United States has become vulnerable to Soviet nuclear attack, to an Arab oil boycott, and is subject to a seizure of hostages by a small power. It has to look after its own interests without forgetting those of the system, and it has to appeal to its allies for help and solidarity. But in all these dimensions, Europe (and Japan) have long been vulnerable and are still more so than the United States. They would suffer more from a nuclear war or from an oil boycott, and Chancellor Schmidt has been quoted as saying that while President Carter had to worry about 52 hostages in Iran, *he* had to worry about 17 million hostages in East Germany.

Their interests are even more at stake than those of the United

States, as Americans never tire of pointing out, but, by the same token, the risks they run are also greater. Yet it is the United States which has the decisive role in dealing with the Soviet Union, with the Arab World, or with Iran. The Europeans are, then, in the particularly uncomfortable situation of being more exposed than the United States to the consequences of failure by the same United States in the management of the various crises in which it is engaged. This makes them vulnerable to blackmail from every side.

The Federal Republic, in particular, is the Western country which has the greatest stake both in Soviet and in American goodwill: because of the achievements of Ostpolitik in terms of human and economic relations with the GDR, and because of the exposed situation of Berlin in one case; because of the protection assured to the same Berlin and to West Germany itself by the presence of US troops and nuclear weapons in the other. It is therefore submitted to progressively more awkward dilemmas: for instance, to buy both more Soviet political goodwill and more economic independence from the Arabs, it becomes more vulnerable to Soviet blackmail by its increased reliance on Soviet natural gas. Yet how are the Germans, the Europeans, or, for that matter, anyone else, to establish a rational balance of risks and of costs for various sides and for various contingencies? The most serious aspect of the situation is that such a calculus is made impossible by the particular unreliability or unpredictability of some key actors and by the *general unpredictability* of the international system as such. The crisis of governability seems to make it impossible for some key governments, particularly that of the United States, to commit their countries to any complex negotiated package or to any coherent course of action. More important still, the West and, more generally, the developed world, have become crucially dependent upon countries which either are to be seen more as anarchical societies than as organised states, or hold values and attitudes which make communication with Western interlocutors utterly meaningless or misleading. Today's Iran is, of course, the most blatant example of both syndromes – but by no means the only country whose next regime is unpredictable except to the extent that one can predict its behaviour to be unpredictable itself.

Perhaps the only valid generalisation about today's international system is its combination of *interconnection* and *heterogeneity*. Everything is related to everything else but no one relates to anybody. Societies are dependent on other societies with which they cannot establish a viable dialogue even on pragmatic or cynical grounds. Functional structures and regional balances do exist and display a certain recognisable logic, but this logic is constantly disturbed by the interference of other dimensions and regions. Each may have its own rationality but the effect of their interaction is as likely as not to be highly irrational.

The conclusion to be drawn from this can only be both pessimistic and relativistic. It is obvious that the world is too discontinuous, too asymmetrical, too universally but unequally vulnerable, and, finally, too unpredictable, whether through arms control or through the co-operative management of interdependence. It is also obvious that there is too much continuity and interconnection for islands of functional or regional stability or rationality to function in isolation for very long. Yet the more troubled the context, the more necessary it is to inject into it whatever measure of predictability can be attained. How to build policies of stability which are not blind to change, of peace which are not blind to conflict, of reciprocity which are not blind to asymmetries of power and purpose, of arms-control which are not blind to politics; this challenge is likely to keep both sides of the Atlantic busy for a long time to come.

If it is to be met at all, each side must learn to appreciate the last of the above-mentioned dilemmas which it has been trying to evade. Americans can neither pursue a purely technical conception of arms control (as was the temptation of earlier administrations), nor (as may be the temptation of the new one), discount the discontinuities which make other areas of the world, including their own allies, difficult to predict and to control, and unwilling to sacrifice their own priorities based on the specifics of their own vulnerable situations. Yet Europeans will have to learn that a political conception of arms control cannot mean that arms control becomes a fig-leaf for subordinating defence to detente, or a sop to the politics of the left-wings of social-democratic parties. To the double dilemma that neither their defence interests nor their detente interests can be either completely separated from, or completely identified with, those of the United States, they will have to come up with a better solution than a complete reliance on the United States for their defence and a complete decoupling from it for their regional detente. How to find the right proportion of continuity and divisibility both in defence and in detente, both in confrontation and in negotiation, both with the East and with the South, between partners, between regions and between issues; this is the hardest yet most unavoidable problem which the 1980s raise for the asymmetrical Western Alliance.

Notes

1. Pierre Hassner, 'Intra-alliance Diversities and Challenges: NATO in an Age of Hot-Peace', Brussels, 1-3 September 1979. Published in Kenneth Myers (ed.), *NATO: The Next Thirty Years* (Croom Helm, London, 1980) pp. 373-97.
2. Irving Kristol, 'Does NATO Exist?' in ibid., pp. 361-73.
3. Robert Legvold, 'Containment Without Confrontation', *Foreign Policy* (Fall 1980).
4. See the evidence in Werner Kaltefleiter, 'Public Support for NATO in Europe', in Myers (ed.), *NATO: The Next Thirty Years*, pp. 397-410 and in Pierre Hassner, 'Western European Perceptions of the Soviet Union', *Daedalus* (Winter 1979), pp. 113-51. Also, 'Les Francais et Le Bloc Socialiste', *Le Figaro*, 12 November 1980.
5. See, for the United States, Samuel Huntington, 'American Foreign Policy: The Changing Universe' in Myers (ed.), *NATO: The Next Thirty Years*, pp. 242-4; and, for France, 'Sondage: L'Armee Rouge en France, 63% des Francais Lachent les Etats-Unis et S'arrangent avec les Russes', *Actuel* (January 1980).
6. Henry Kissinger, 'The Future of NATO' in Myers (ed.), *NATO: The Next Thirty Years*, pp. 3-21.
7. Albert Wohlstetter, 'Illusions of Distance', *Foreign Affairs*, (January 1968). See my discussion in 'The Nation-State in the Nuclear Age', *Survey* (April 1968).
8. Colin S. Gray, *The Geopolitics of the Nuclear Age* (Croom Helm, London, 1977).
9. Albert Wohlstetter, 'Threats and Promises of Peace', *Orbis* (Winter 1974). Wohlstetter coins the term 'firebridge' to account for this ambiguity.
10. McGeorge Bundy, 'Strategic Deterrence Thirty Years Later: What Has Changed?' in *The Future of Strategic Deterrence, Adelphi Papers* no. 161 (IISS, London, 1980), pp. 5-12.
11. See Robin Ranger, *Arms and Politics, 1958-1978: Arms Control in a Changing Political Context* (Macmillan, Toronto, 1979).
12. Francois de Rose, 'European Concerns and SALT III', *Survival* (September-October 1979).

10 EAST-WEST RELATIONS AND THE POLITICS OF SECURITY[1]

Helmut Sonnenfeldt

It has now approached the status of truism to state that the outlook for US-Soviet relations, and East-West relations generally, is, at best, far from promising. However, this consensus, although perhaps trite, is none the less of fundamental importance for the future of Western security, demanding vigorous political and military leadership in order to reverse, or at least manage, the adverse secular and political trends which it reflects. In this connection, it must be born in mind that the situation in which the West finds itself as it enters the 1980s is only partly the result of *inherent* contradictions and intractable problems in managing East-West relations; it is, as much as anything else, the result of missed opportunities, misguided expectations, and distorted goals. What follows is an attempt to sketch some of the lessons to be derived from the experience of the 1970s in the hope that they might guide us in the 1980s.

The effort mounted by the United States and its allies in the early years of the 1970s constituted an attempt to define and develop certain ground rules for the conduct of Superpower relations in an era of structurally inevitable competition, increasing and increasingly uncontrollable instability throughout the world, and burgeoning and progressively refined and diversified military forces. The first tentative steps in the development of these rules of engagement, initiated by the French and Germans in the 1960s, and followed by the United States in 1972 and 1973, were valuable expressions of how a somewhat less dangerous world might function. Those principles, subject to subsequent misinterpretation and malignment, largely a result of misplaced and overblown expectations and hopes, were never intended to assume an autonomous guiding role simply through their expression on paper. East-West agreements of this sort cannot be self-enforcing, and should not be thought of as such. More importantly, they should not be portrayed to the bodies politic as fundamentally transforming. Such rules, if they are to function as such, must have as a foundation a structure of incentives and disincentives which, as with all successful guidelines, would make those ostensibly subject to them see more advantage in observance than violation. The structure of risk and benefit erected during the 1970s was inadequate for this task.

Specifically, the institution of arms control was subject to excessive hopes related to its effectiveness in restraining Superpower behaviour in the 1970s. It cannot be argued credibly that there does not exist a shared fear of nuclear war, and a joint interest in its prevention. It does not require a profound, almost anthropological, understanding of Soviet culture to accept the proposition that a nuclear war is not a top priority in the Soviet Union. What *is* open to question is whether this common fear has meaningful implications for practical realms of strategic conduct and political co-operation which would suggest the coincidence of interests and policies for Superpowers who operate more in spite of, than according to, this joint recognition. Indeed, it could well be argued that far from generating pressures for joint management within a peaceful consensus, the fear of nuclear war provides unique opportunities for manipulation, and can, in some instances, intensify crises, rather than reduce them on the assumption that the fear of the other side will be the governor in the particular crisis, and will thereby lead to its defusion.

In the realm of arms control, this assumption of a 'shared fear of war' has hardly proven to be a particularly helpful or productive catalyst. Moreover, it is even questionable whether arms control negotiations themselves have had very much to do with the prevention of war. Indeed arguments can be made that some of those agreements and negotiations might, conceivably, have increased the dangers rather than reduced them. (Thus, the ABM treaty, useful in its day, may, if unchanged over time, preclude programmes, such as point defence, which could contribute to the protection of ICBMs, command centres, etc.) The argument made in the American SALT debates that the agreement should be ratified because without it we run the risk of nuclear war, and with it we stand a better chance of preventing a nuclear war, is spurious. Yet this expectation, this calculus, was one of the illusions of the 1970s, contributing to the failure to take realistic steps to construct a lasting framework of reward and risk.

A further element in the mythology of detente which informed the 1970s was the assumption that greater economic interaction between East and West could have a restraining and stabilising influence on the political and military behaviour of the Soviet Union. This particular proposition has been untested because, on the part of the US, the policies which needed to be pursued with respect to economic relations remained unexplored, because, paradoxically, legislation passed in the American Congress tied economic relations to Soviet immigration practices, and denied the American Government the instrumentalities for conducting a coherent economic policy toward the Soviet Union.

Economic policy and economic interaction can play a constructive role in the containment of antagonism and in the promotion of restraint, but it can do so only if it is part of a broader strategy, rather than simply a series of micro-decisions made by individual firms whose function it is to do business. The firm is not in the business of setting national strategy. This has to be a matter of national decision-making, of national consensus by national authorities, and it is extraordinarily difficult in our political systems to subordinate economic policies to centrally conceived and centrally agreed political strategies. However, because governments, including that of the United States, have failed to view economic policy as part of the strategy of managing conflict, it has turned out to be the case that the Soviets have been far less restrained by the possibilities of either *greater* economic benefits in their relations to the outside world, or the danger of having economic benefits curtailed, than Western countries have been inhibited in responding to Soviet lack of restraint. Once again, this has been the result of an inadequate structure of incentives. The inflexibility and insulation of processes of economic interaction from political control has not generated a supple instrument for the manipulation of benefit as a response to Soviet behaviour patterns. In the present situation, particularly since the invasion of Afghanistan, this phenomenon has been very much in evidence in parts of Europe. Western European countries have become excessively dependent upon economic relationships with the East and in particular with the Soviet Union. The reluctance to tamper with economic relationships has been a substantial impediment to the management of conflict.

At a more general level, there has been an endemic problem arising from differing definitions of the notion of detente. This is not merely a matter of definition, for the use of similar terms serves both to disguise differences in ends and means and to generate a false climate of consensus which then breaks down all the more severely when that consensus is exposed as ephemeral. Detente has been conceived both as a permanent condition – either in being or imminent – rather than as a complex political process. Moreover, it has been defined in terms which imply the lack of conflict, ignoring the fact that East-West relations can at times benefit by actions that might raise tensions rather than reduce them, through the clarification of interests, the demonstration of wills, and the definition of the boundaries of what is tolerable.

It may well have been that in the early 1970s, when American-Soviet relations began to intensify, the Soviets assumed that the recognition implied in the agreements made at that time – a recognition of their role as a major power and as a world power – allowed them far greater scope for expanding their world role than the Americans and others intended. These early efforts to work out the relationship with the Soviet Union coincided with a time of American retrenchment and the debate about the US role in the world and the vices of intervention and covert operations. It may well have been in this important instance that the Soviets took as a permanent condition a degree of American tolerance for Soviet expansion that was in fact temporary. In any event, since at least 1975 and the events in Angola, there has been a notable swing in American opinion, not to rampant interventionism, but certainly to a profound unease about the extent to which Soviet influence and presence have been expanding in the world.

The Soviets claim that in this respect they were seriously misled. Certainly, the Soviet Union had a genuine sense of entitlement to a larger world role, and resented the notion that someone should attempt to place limitations upon that. In this regard, the Western rapprochment with China, which got its major momentum, after the pioneering work of the Europeans and Canadians, in 1971 and 1972, encouraged the Soviets to make greater efforts than hitherto to prevent and counteract a possible encirclement by a hostile coalition built around Japan, China, Western Europe and the US, and led them to a strategy already there in embryonic form when Brezhnev took over from Khrushchev: a conscious strategy of seeking the encirclement of China. This, by Soviet definition, was a legitimate response to threats seen from the outside world and to violations of the understandings embodied in the statement of principles and ground rules for conduct that were written in the early 1970s. That this was so is a reflection both of the inevitable misinterpretation of those rules between East and West, and within the West, but more importantly of an inadequate understanding of the political and geopolitical interests of the systems involved.

Furthermore, the European stake in detente has come to be increasingly greater than that of America. For the Europeans, in particular the Germans, detente has had practical and human consequences that simply have not been operative in the case of the US. In addition, there has been a growing divergence in the public moods on the two sides of the Atlantic with respect to the phenomenon of an expansive Soviet Union. On the whole, in the US, this has led to disenchantment, a certain militancy – to a public attitude of saying somewhere, somehow we have to assert our *own* interests. It has led, in the US, to a marked change in public and congressional attitudes, finally reflected in more affirmative Carter Administration actions with regard to defence expenditures. This manifested itself for the first time in 1976 when it proved possible to have the first real increase in defence expenditure in the US for some time. In Europe – and here obviously one has to allow for many differences – while it has proved possible to achieve certain increases in defence expenditure, the reaction to evidence of Soviet expansionism on the whole has been to urge a redoubling of efforts to negotiate, and to preserve the gains of detente. Tensions in one part of the world are not to be imported into the European theatre. This divergence of European and American public attitudes ought not to be minimised. They are more significant and more troubling in terms of managing West-West relations than some oversight in consultations. These tides in public attitudes are much less subject to 'fixing' through improved instrumentalities and through improved methodologies. We are here confronted with attitudinal issues and problems in the Western community that are much more serious to cope with.

Pervading all these trends, however, there is simply no way of getting around the cumulative effect of the Soviet military programmes of enormous and sustained momentum that we have witnessed over the last generation. These programmes have their roots in the Stalinist period and the period at the end of the Second World War. Nevertheless, their particular effects have become especially noticeable in the last ten years. It is not adequately recognised that in years in which American defence efforts fluctuated greatly and were diverted in enormous amounts to Vietnam, Soviet defence expenditures increased somewhere between 3 and 5 per cent in real terms, doubling in the Brezhnev era. That of the United States not only did not double but steadily declined, and has yet to reach the level of 1963 in real terms. Much the same can be said about the defence budget situations in other Western countries. This has got to make a difference at some point. It will not do for an American spokesman to tell us that effective parity in strategic weapons has really existed since the late 1950s, and that American leaders have been fully conscious of this fact. This is untenable unless one stretches the concept of parity to include what was by all conceivable indices a very substantial American advantage in strategic weaponry in the late 1950s and early 1960s, and what amounts now, by the same indices, if not to an American disadvantage, certainly to no substantial American advantage, and probably to some very substantial American vulnerabilities. To equate the two situations, or to treat them as strategically insignificant, is to defy strategic reality. Relative strengths have real political, if not military implications; relative strengths which have very real implications for the manipulation of crises through the medium of advantages in the escalation process are even more serious. It is unnecessary here to enter into an argument about the significance of counterforce threats to American land-based systems, threats to American command and control for sea-based and other systems and various other debilitating trends effecting the efficacy of the American strategic forces. However, if one compares the late 1950s, when in the view of some the age of parity was supposed to have begun, with the early 1980s, there is a serious difference in strategic effect between a capacity to destroy American cities and the capacity to destroy American weapons, placing the use of American weapons in some question and raising very much the issue of launch on warning doctrines. And this, of course, has serious implications for the credibility of the American guarantee in Europe – both within Europe and the United States itself.

At the regional level in Europe, the massive improvements in Soviet conventional forces have been steady, and have changed the situation substantially to the advantage of the Soviet Union. Perhaps more important, though obscured by the LRTNF debate, is the heavy nuclearisation of Soviet forces in Europe. This is not simply a matter of the SS-20 but of whole families of nuclear weapons in large numbers that are continuing to be deployed with Warsaw Pact forces, and which permit the Soviets to contemplate a combined arms strategy in *any* possible war in Europe. It is this development that makes the TNF modernisation issue so very important for the West. It is unfortunate that it has become tied so closely to the development of the SS-20 and the Backfire, and that the arms control initiatives that the West found it necessary to associate with its decision to produce and to deploy its own long-range systems, were tied exclusively to the Soviet deployment of the SS-20. In any event, the nuclearisation of Soviet forces in Eastern Europe and in the Western military districts of the USSR has added serious new questions to the pre-existing ones about NATO strategy, at least until the time when Western capabilities – conventional and nuclear – have been improved in line with the decisions that have been taken. Not the least of these is resolving the problems presented by a NATO strategy of planned insufficiency in regional balances under the rubric of US central superiority, when that superiority has evaporated.

At the level of regional conflict, it needs to be stated bluntly that the Soviets have, through the invasion of Afghanistan, opened a new front in the Near East. Whatever the motivations of the Soviet invasion of Afghanistan, and whatever the further course of events there, the *effect* is to place Soviet forces now at the two corners of the Near East – at the south-west corner in the Saudi Arabian peninsula and in the Horn of Africa, on the one hand, and at the north-east corner in South West Asia, on the other. Afghanistan, even though it may be seen as an extension and mutation of the Brezhnev doctrine, is dissimilar from Hungary or Czechoslovakia because there is no power balance in the area. There was after all a North Atlantic Alliance opposite Czechoslovakia and Hungary, whatever the flaws and the shortcomings in its military posture. There is nothing like that opposite or adjacent to Afghanistan, and here is the connection to the fall of the Iranian monarchy, and the internecine struggles in the Persian Gulf area. The region adjacent to Afghanistan is highly unstable in terms both of the situation within countries and relationships between and among countries. The lack of a stable power balance in this area, combined with its critical economic and geopolitical importance represents a challenge and impediment to the global stabilisation of relations between East and West, and threatens – in reverse manner to traditional patterns – to feed back into the process of detente in Europe. The most immediate problem is the management of an economically vital area presently without commensurate Western political influence and military resources.

This leads to another aspect of the military problem: the growth of Soviet naval power and the worldwide presence that has become associated with it. Arguments about master plans or opportunism are essentially senseless. What we have seen is *a pattern*; whether it was intended or inadvertent is an open question. Power has a way of creating its own opportunities. The combination of Soviet power can be asserted at different places and different times throughout the world. This is not to suggest that the Soviet presence that has been projected, either directly or through proxies, is necessarily immutable. There will continue to be instances where it will be ejected for one reason or another, as in West Africa and Egypt. Nevertheless, it is a characteristic of Soviet behaviour that the effort never ceases, while the opportunities will never cease to exist. The US, as a maritime power that must project force over great distances when it needs to act militarily, can no longer count on what it has been able to count on since the Battle of the Atlantic: it can no longer expect to engage in military operations on the Eurasian land-mass with totally secure lines of communications. It was possible in Korea, Lebanon, Vietnam and elsewhere. Currently, however, the traditional problem, of military defence against Soviet pressures – at least on the Eurasian land-mass – is infinitely more complex than it was in the past. The West learned in the 1940s and 1950s that a distant maritime power that confronts a power on the Eurasian land-mass must have allies on land. That is not a condition that exists at present in the region stretching from Turkey, Israel and Egypt in the West, to the East. We must, therefore, take with the utmost seriousness what has happened in Afghanistan. Without making any predictions concerning Soviet intentions, or further Soviet action, it is a relatively objective observation that there is a problem of massive proportions in the lack of stable or strong regional allies. To this somewhat melancholy and worrisome catalogue should be added the fact that the 1980s are likely to see more urgent Soviet economic problems than heretofore, in particular energy, and it is therefore quite possible that what is happening in the Near East is, among other reasons, intended to serve as a means for determining the terms upon which the Soviets, on behalf of COMECON, will become a net importer of oil in the 1980s. With the Soviets lacking the hard currency that they are now earning through the export of oil they will be profoundly interested in the terms by which oil can be imported.

In response to these past and imminent developments, a redress of the central and peripheral military balance is a necessity which must not be sacrificed to the fulfilment of other social obligations. It is a *sine qua non* for the conduct of policies dealing with the Soviet Union and with security issues between East and West generally. It is not, as we know, a problem *exclusively* for the US, but for the Allies and others. It is difficult to see how we can live out the millennium with the Europeans seeing themselves as having vital interests in areas outside Europe – mostly economic ones – while *at the same time*, by and large, leaving the physical defence of those areas to the US, whose reliability and whose skill is questioned in many quarters in Europe, and whose willingness to act on behalf of Europe is increasingly under attack. Europeans, therefore, must come to grips in some way with how they are to contribute to the defence of their vital interests over and beyond diplomatic skill and the efficacy of their economic policies. This is not a matter of extending the jurisdiction of the North Atlantic Treaty. But while it is quite true that the North Atlantic Treaty is confined in its effect and applicability to a particular geographic region, the interests of the allies are not. The Alliance will not long endure with integrity if threats to the security and well-being of allies that emanate from outside the region of the commitment of the Alliance are left unattended and are not made a subject of concerted action.

We will have to face the fact that there will be, in the 1980s and beyond, the necessity to recognise Soviet interests, and, indeed, an expanding Soviet role in the world. But what we need to be clear about, as individual governments and bodies politic, is that we should not accept Soviet efforts to establish *exclusive* positions of influence and *exclusive* roles in particular parts of the world. We must make clear our capacity and our readiness to resist such efforts. This isn't simply a matter of that elusive notion of will, although that is involved. It is a matter of hammering out a political consensus as to what interests are to be defended where, and in what manner.

This is not something that can come exclusively from the grassroots of the Western electorates. Although there is increasingly through the West – although the extent of this movement is significantly varied in its magnitude – a popular inclination to redefine and put on a more stalwart basis our relations with the Soviet Union, this is a function of political leadership and creative statesmanship.

It will be necessary and desirable to proceed with arms control negotiations, both for whatever narrow strategic effect such agreements might produce and for the political and psychological domestic leverage this will provide for more basic remedial steps which must be taken at the military level. This is a political fact of life which must be accepted, though not pandered to in a blind and inflammatory manner which will block policies at other levels which are vital adjuncts to arms control in the creation of the feasible and productive set of incentives outlined earlier. But we must disabuse ourselves of the notion that arms control negotiations and possible agreements will materially assist us in dealing with the military problems that we face. It is possible, perhaps, to derive some marginal benefits with respect to certain types of weapons – the pace of deployment or certain characteristics – but it is open to question whether the notions of stability that originated in the 1950s and 1960s, and the negotiating efforts which they generated, can give us much sustenance in the 1980s and 1990s.

Beyond greater realism about arms control, we must find a more co-ordinated set of approaches to the formulation and conduct of economic policy. We are in a period when the Soviet Union, due to the shortcomings of its own economic system and those of the East European countries, has come out into the world to seek the benefits of the international economic system. The decisions in this regard cannot be left to the individual entrepreneur. Western governments and common institutions have got to find some better way to deal with this issue. We cannot continue the incongruous situation where we are in a deepening conflict on issues of security interest while economic interactions and reliances increase. There must surely be some way in which governments can bring economic relationships into some degree of correspondence with a state of political and security relationships, and there must surely be a way in which Soviet leaders, especially the ones that are waiting in the wings, can be brought to understand that if one seeks the benefits of the international economic system one must also contribute to the discipline of the international order as a whole.

These are glowing generalities, but it is impossible to avoid the conclusion that we have done badly – individually and collectively. We must do better particularly because we can expect that in the next three to five years, if not before, the Soviets will enter the international energy market as net buyers. How is that going to be financed? Are Western lending institutions, financial institutions and Euro-currency markets going to finance the Soviet pressure on the international oil market that will raise the prices for the rest of us? Are we going to be in a situation where the Soviets will index their arms and their other limited manufactures to the price of oil, and will impose barter deals on oil producers that are highly disadvantageous, but which they cannot avoid because of the political condition in which they find themselves?

These are important and crucial issues; it is past time for Western governments and collective institutions to get to work on them.

Note

1. This chapter represents an edited text of informal remarks made at the 1980 Millennium Conference.

11 MILITARY POWER AND ARMS CONTROL: TOWARDS A REASSESSMENT

Hugh Macdonald

Introduction

The argument of this chapter is ultimately simple: arms control in the Alliance is a matter of political will concerning military power and its place in the international relations of the developed world. As such, it cannot be examined in isolation from considerations bearing on the role of, and attitudes to, military power itself. Presently, arms control is secondary in importance and ambiguous in purpose, because the will to solve problems of military power has been absent. The developed world as a whole has chosen to accept and maintain what is here termed an Evolved Cold War structure of deterrence, which entails continuous scientific weapons development and high levels of defence expenditure, in which arms control functions as a palliative, but not a solvent, of East-West tensions and Alliance problems. This view is not an optimistic one: in the face of the risks of moving to a different arrangement of international relations, the will to take these risks has been absent.

The underlying patterns of the use of force in the postwar world are symptomatic of a special and in certain ways isolated structure of developed world international relations. This suggests that functional forms of arms control developed to support deterrence cannot be transferred to developing world problems, and yet there is a dynamic relationship between conflict in the developing world and the continuance and stability of the Evolved Cold War in the developed world. The problems and risks of military power thus increasingly involve divergent interests, admixtures of developed and developing world crises, fluctuating Superpower tensions, and defence-based Cold War rhetoric.

With so much that is imponderable, it seems better to seek to analyse the past and the present than to speculate or prescribe for the future. Accordingly, the attempt to describe some problems of military power and the existing dilemmas of arms control is followed only by an attempt to analyse some elements that have contributed to this: an 'international consciousness' of competition and reliance upon scientific weapons development to manage the Evolved Cold War; an ambiguous perspective upon the use of force; an ambivalent notion of rationality; and a conservative philosophy of arms control. If change is to follow, then it will need to account for these strong roots from the past which have entrenched themselves in our will, and institutionalised themselves in our Alliance.

There is a paradox here. The present structure of military power in the developed world leads NATO to seek arms control solutions to problems of military power, but the existing status of arms control institutions – particularly SALT and MBFR – frustrates this, because it *supports* continuing military competition. In turn this includes the diffusion of weapons, technologies and ideologies from the developed to the developing worlds. Moreover the Alliance – and in particular the United States – tends to blame the Soviet Union for these circumstances. Attempts are being made to solve NATO's conflict of interests and institutional inflexibility by extending its ambit and the scope of intra-alliance political consultation. It has begun to seek regional and local solutions in terms of military containment. This seems bound to further stretch its resources and its cohesion.

Yet there remains the dilemma of recognisable feedback effects of competition and crisis upon an isolated and stretched Soviet regime, which might clumsily resort to force in areas of delicate East-West engagement, such as Poland. So long as the Soviet Union adamantly clings to its Superpower status as a guarantee of its revolutionary identity elsewhere, and this identity, evident only in incendiary and usually unsuccessful attempts to undermine Western interests in the developing world, is used to legitimate a dangerous misallocation of resources at home, it remains condemned to a pre-modern economic and social structure. This indicates the intangible connections between domestic politics and foreign policy, and between international relations in the developed and developing worlds. The West has more scope, and perhaps more responsibility, for rendering these intangibles into effective linkages, because it is inherently more open and economically stronger, but it is impossible to do this within an Evolved Cold War framework which plays upon the West's greatest weakness, namely the peacetime threat and use of military power.

Uncertainty over the future of arms control cannot therefore be resolved by prescriptions for arms control negotiations. Such prescriptions depend entirely upon the evolution of developed world international politics, and particularly Superpower relations. In turn, however, Superpower relations depend peculiarly upon the way in which the domestic politics of foreign policy develop in Washington and Moscow. Each side is capable of influencing the other, but it is peculiarly difficult at present to guess how these influences will work. A global American strategy resting upon an undifferentiated *realpolitik* may succeed in containing perceived instabilities in the Middle East, the Gulf, Africa and Latin America, but may fail to preserve detente in Europe. On the other hand it is conceivable that a signalled return to Cold War may prompt the Soviet Union to concede more to detente, including arms control. Much depends upon the subtlety, the conceptual refinement and the geopolitical discrimination which is induced in Washington. Similarly, the Soviet Union could simply conclude that two must play the game of spheres of influence. This would not be particularly difficult because of the Soviet Union's strengths: a transcontinental position which allows it cheaply to bring great pressure to bear upon many neighbours, its revolutionary identity which rationalises incendiary behaviour in the developing world, and its growing capacity to project military power over great distances. On the other hand, President Brezhnev demonstrably prefers detente and the possibility of resolving some of Moscow's great economic and foreign policy problems; and it may be, if a more consistent Western policy of carrots and sticks is developed, that considerable arms control benefits are possible. But prognostication carries us away from the task initially posed: a diagnosis of the nature and meaning of arms control in its connections with the contemporary context of military power. An examination of that context is, therefore, a necessary preliminary.

The Context of Conflict

The international system within which arms control agreements have functioned for the last 35 years was established at the end of the Second World War. Its features might be enumerated in a number of ways, but for present purposes Geoffrey Barraclough's words will do:

> the changed position of Europe in the world, the emergence of the United States and the Soviet Union as 'super powers', the breakdown (or transformation) of old imperialisms, British, French, Dutch, the readjustment of relations between white and coloured people, the strategic or thermo-nuclear revolution.[1]

In conjunction with these phenomena, several secular patterns have emerged: the growing impact of economic and resource questions; the shift of axis from East-West to North-South politics; the emergence of a large number of independent power centres; the creation of novel forms of international 'issue management' such as the Law of the Sea Conference and forums for dialogue (and monologue) about a New Economic Order; the foreign policy burden which has devolved upon one country – the United States – in sustaining the existing order; and the proliferation of nuclear and conventional war-making potential, and of low level international and domestic violence. No such list establishes an argument either way concerning the relationship of pattern change to structural change and thence to conflict; it does constitute evidence that this kind of relationship may be crucial.

Patterns of use of force in the period since 1945 can be divided into two periods. These periods may be called the Cold War, and the Evolved Cold War. A typology of uses of force suggests a transition from the former to the latter at the beginning of the 1960s, and a correlation between the overall structure of international relations and forms that organised violence takes. In the Cold War there were numerous Superpower confrontations, between January 1946 and October 1962. None of these confrontations resulted in direct violence between the two adversaries. Such 'deterrence' behaviour contributed to stabilisation of the Cold War. However, the Superpowers became increasingly insensitive to interventions in other parts of the world. Countries under colonial rule and seeking independence became 'power vacuums' between the hard lines of demarcation of the two blocs. Colonial powers allied to the United States resisted national liberation movements, either for reasons of prestige or as a function of Cold War alignments. Aspirant 'nations' drew upon the ideological cleavage between the Superpowers. In turn, the Superpowers and other great powers from time to time made forceful excursions into the 'underdeveloped world'. Neutralist forces were unable to affect significantly the structure of the international system. For the most part, wars fought between developing countries were of a small scale or had overt military support from one or other Cold War bloc.

The Evolved Cold War – a changing pattern of relationships clearly within the same post-war system – has been marked by significant changes in the use of force. First, direct Superpower confrontation in Europe has ceased, and given way to close consultation in an arms control framework. Secondly, wars of national liberation have decreased in frequency with decolonisation and the creation of a very large number of new states, but wars between developing countries have increased in scale and frequency. The first example of an exclusively 'Third World' interstate war on what threatened to be a large scale –

but in fact was not – was the Sino-Indian war of 1962. It was followed by the Indo-Pakistan Wars of 1965 and 1971, the third and fourth Arab-Israeli wars of 1967 and 1973, the Ethiopia-Somalia and Libyan-Egyptian wars, and latterly by the Sino-Vietnamese, Tanzania-Uganda and Iraq-Iran wars. All of these wars have been fought between relatively newly independent states for 'intermediate' objectives rather than for 'fundamentals' such as national independence or survival. In a sense these have been wars of readjustment, much more akin to the classical pattern of political use of force analysed by Clausewitz.

A third variant in the pattern of use of force has been the spread of domestic violence, and covert rather than overt interventions in it by outside powers, including the great powers and such new interventionaries as Israel, Cuba, Vietnam and South Africa. Civil Wars in Nigeria, Angola, Lebanon, Zimbabwe and South East Asia provide examples, but domestic violence has grown in the developed world as well, with terrorism, urban guerilla groups and civil strife affecting several European countries and Japan.

These patterns suggest trends in the use of force rather than hard and fast categories. For example, the Korean war and the Indonesia-Malaysia confrontation were large wars between developing countries, though supported by one or more of the world's main military powers. The Vietnam War and extensions of it into other parts of South East Asia provided examples of protracted, overt, Superpower intervention against social and political change in that region after the Evolved Cold War had begun. None the less the general trends seem clear: no wars in the developed world; wars of adjustment in the developing world; and domestic violence everywhere, encouraged by the continuance of Superpower conflict. It can be inferred from this that the pattern of use of force in the international system follows the overall pattern of political relationships. The different scale of violence between developed world and developing world conflicts can be accounted for by the difference between a highly structured context of developed world international relationships, and a relatively unstructured context of developing world international relationships. Furthermore, domestic social change appears to be as significant an agency as ideology or state policies in providing the occasion and locus of violence.

In the post-war developed world the pattern of use of force has been modified, in degree at least, by two structural factors. The first is the type of economic interdependence shared by almost all developed states. The centralisation of states in their domestic contexts has been accompanied by the acceptance of horizontal linkages across national economies. This has helped produce a perception of 'interests' which discounts the use of force in relations between developed countries.

The second factor is the kind of uncertainty imposed upon the political use of force by the existence of nuclear weapons. Most importantly, this factor has inhibited the use of force between the blocs – the Superpowers and their allies – where inhibitions due to economic interdependence have not applied. These structural modifications of the politics of force have worked largely by transforming issues of change in the East-West context, to which force *might* have been applied, into complex political dialectics between the Superpowers and their allies. SALT, the Ostpolitik and Sino-American relations over Taiwan and the offshore islands remain its most remarkable achievements to date.

But complex and subtle though these processes are, it remains true that an ideological cleavage of uncertain future consequence still divides West and East; that the dialectics of change have also entailed the suppression of change, sometimes, as in Eastern Europe, through the tacit agreement of both Superpowers; and that change without war in Europe and between the Superpowers has not obtained in other parts of the world, where violence has often been due to, or exacerbated by, this East-West framework. This suggests the following additional inferences within the East-West context: although a relationship can be discerned between arms control institutions and underlying conflicts, this relationship cannot be taken out of context; it depends heavily upon wider structural factors that are distinct attributes of developed world international relationships; and if arms control is to apply to the developing world, new institutions and processes will be called for.

The rest of the world does not accept the terms of the East-West stable confrontation. This has been becoming ever clearer at the level of developing societies for many years. Increasingly, it appears, governments of developing countries are adopting East-West alignments only to the extent that so doing furthers their own relatively short-run domestic or foreign purposes. This is scarcely surprising as the clashing conceptions of Soviet Communism and American Democracy are not replicated exactly anywhere else in the world. Though this does not necessarily mean that there is no worldwide ideological struggle, it does mean that it no longer has any prescribed political form; as the distribution of power and membership in the international system changes, strategic political choices, including judgements about military power and arms control, cannot be based on *a priori* ideological or political judgements.

It must still be a subjective judgement how far the rejection of the

East-West framework will itself become a general 'cause', and how far conflict will be due to attempts to displace it by some other set of prevailing norms. The failure of the Non-Aligned movement to exert effective collective pressure upon the Superpowers, and the isolated and vastly different radicalism of, say, Libya or Iran, suggest that neither anti-colonialism nor positive revolutionary ideologies can cohere sufficiently to disrupt the structure of developed world economic, political, or military relationships. But recent past evidence, for example that surrounding global energy issues since 1973-4, suggests that the raw material of conflict and misperception is abundantly available in differing demands upon the international order, especially in development issues, and that it may be strengthened or triggered by social transformation within, and by the diffusion of independent power among, societies and states. It also suggests that there remains both the hope of solutions within international frameworks, and the appalling possibility that these too may become sources of conflict and, sooner or later, violence.

States and groups beyond the blocs may be co-opted by the terms of Superpower conflict, and developed world international relationships, due largely to 'parallel choices'. But these correspondences will persist only for as long as the developed world is powerful enough, or willing enough, to offer them (or pay for them) as terms of reference; and as long as other states are so lacking in power or room for manoeuvre that they do not desist.

There is, then, a fundamental difference between the developed world participants in blocs, for which the overarching framework offers positive security, and developing countries for which it does so only negatively. This difference will become clearer and more important as the voluntarily neutralised politics of force of the East-West relationship meets the aspirations of revisionist new nations which will have acquired the substance as well as the symbols of independent power, and thereby the incentive, and probably the urge, to try to transform negative into positive choices. This has been recently and strikingly exemplified by events in Iran, where the former Shah's parallel choice of close alignment with the United States has been violently rejected.

Complex local instabilities may again draw the two Superpowers into renewed confrontation as they did in October 1973. Africa has re-emerged as an arena of conflict for powers which cannot fight each other directly in Europe. The Indian Ocean, the Gulf, and to a lesser extent the Mediterranean are areas of severe competition, where conflicts in contiguous states invite the Superpowers to competitive naval demonstrations (Bangladesh 1971, Iran 1980) and to overt or clandestine interventions (Lebanon 1976, Ethiopia-Somalia 1977, Yemen 1978).

More recently, Soviet intervention in Afghanistan has led to a resurgence of Containment thinking in Washington, to serious planning for a Rapid Deployment Force, and to NATO-wide consultation about extending alliance interests and commitments to strategically sensitive areas such as the Persian Gulf. But it is important to note that every major outpost of Containment in the developing world, from Algeria to Vietnam, has fallen to revolutionary-nationalist rejection of suffocating foreign policy alignment with the United States. Renascent Containment is no more likely to succeed by employing the fig-leaf of NATO-wide interests, for in any guise Containment, in its emphasis upon military power, is ubiquitously insensitive to domestic change in developing societies. Moreover those recent issues which have brought Superpower confrontations have also brought deep disagreements over crisis management within NATO.

One vital source for disagreement within NATO is that it remains problematic what effect the spreading and ever more bloody domestic conflicts, and Superpower interventions in them, will have on the stable East-West confrontation, particularly because of growing capabilities for the exercise of conventional military power, and even selective nuclear options, over considerable distances. At best, conflict elsewhere adversely affects the complex balance between confrontation and co-operation that has moved developed world relationships into the Evolved Cold War. Beginning with the Yom Kippur war, Superpower detente and, above all, the ambitions of Americans for linkage at the level of international stability, and of Russians at the level of trade, have been repeatedly shaken by the unavoidably gratuitous effects upon foreign policy making of conflicts beyond the developed world, into which the Superpowers have been ineluctably drawn by a confusion of hopes and fears through which they feel bound to try to discern their interests. At worst there remain not ambiguities but real instabilities within the central strategic balance, the military balance in Europe, and the 'balance of advantage' elsewhere. These coalesce around technologies which make nuclear and conventional weapons more precise, and upon the interaction of such developments on both sides of the East-West divide; once again a Serbian ultimatum – involving the legitimate rights of a developing country and the terms of its development in a wider international order – could drive the great powers into purposeless, immeasurable destruction.

The implications of all this for arms control are profound. For all of the differences of context between developed world and developing world international relationships, the consequences of conflict in one context *do* affect relationships in the other, not uniformly or automatically, but persistently and crucially in cases where the Superpowers cannot be brought to agree, tacitly or explicitly, to seal them off. Clearly, major international policy choices take this into account. Arms control becomes a matter of fundamental choice in something like the following terms. To carry over East-West ideological confrontation into developing countries' conflicts means exacerbated violence, undesirable forms of development, and an increasing probability that developing societies will ultimately reject the stable structure of East-West coexistence. To seek to establish some controls over the use of force and competition which will appy to the relatively unstructured developing world context implies *inter alia* that international conflicts must be dealt with in their own terms; that is by adaptive, constructive, local measures in the first place. But such measures must be linked to diminishing Superpower competition, and to preventing feedback from developing world conflicts to East-West stability. Therefore international conflicts cannot be resolved by arms control proposals and measures intended to gain unilateral advantage in developed world international relationships. This applies as much to West European oil-for-arms deals as it does for blanket American support for Israeli actions as the best counter to Soviet influence, or to Soviet fraternal assistance to change, followed by the maintenance of hopelessly unstable and unpopular regimes in Afghanistan.

The State of Arms and Arms Control

There is almost universal agreement that arms control is in deep difficulty, lacking inspiration to solve present problems and new directions to face future ones. It is felt that SALT has reached an impasse, that even cosmetic adjustment of the accords so painfully reached in mid-1979 would be an admission of failure. The SALT process apparently cannot be backed down the long tunnel it has been in for so long; it cannot be given any new position in international politics. It is felt that arms control in Europe has failed to make political advances because the painfully slow and indirect process of MBFR has been outpaced by dramatic strategic developments in the region. Differences of interest and future development within the NATO alliance appear to make it harder all the time to relate regional arms control to efforts at the central level between the Superpowers. The separate European and world identities of the Soviet Union – and its markedly different modes of politico-military behaviour – seemingly necessitate more clearly differentiated responses than in the past, but 'differentiated distancing' is what the Alliance is above all dedicated to avoid. In all of this there is a plethora of speculation, and almost total uncertainty, about the form and direction which recognition of these factors could take.

Advocates of arms control feel pessimistic because so little has been achieved in reducing levels of military preparedness, military budgets, research and development of new weapons systems, and new patterns of military deployment. Opponents of arms control are pessimistic because detente has not contained manifestations of fundamentally contradictory interests which have been so prominent beyond the limits of the developed world since the October War of 1973. Indeed, arms control has increasingly been blamed for exacerbating these tensions, encouraging a false sense of technical stability, permitting the Soviet Union greater political and military latitude in unlinked areas, and inducing self-deterrence in American handling of confrontations. It is easy to see truth in both views of arms control. But since what is crucial is the nature of American-Soviet competition in a quickly changing international system, it is not easy to confidently reconcile or arbitrate between these views themselves.

Yet many facts are not in dispute. Since Korea, the price for the dichotomy between security in the developed world and conflict everywhere else has been a nuclear arms race between the United States and the Soviet Union. That arms race has been controlled by political self-restraint and by a gradually elaborated technical and institutional structure of deterrence, of which arms control has become a central part. Despite arms control, the terms in which deterrence has evolved have not been 'minimal' and have tended to become 'maximal'. For example, the current SIOP reportedly comprises some forty-odd thousand strategic nuclear targets. The United States has some ten thousand available strategic nuclear delivery vehicles (SNDV). Under the terms of SALT II that total could rise, to perhaps fourteen thousand SNDV. The SIOP does not include many of SACEUR's nuclear targets, and for regional purposes there are up to ten thousand American tactical nuclear warheads. For the United States today, read the Soviet Union tomorrow, and in certain special categories, yesterday. Moreover, nuclear competition has meant stronger rather than weaker conventional defences, at least comparing the pre-Korea years with those that followed. In central Europe today the standing forces of NATO and the Warsaw Pact amount to some two million men. Together with ready reserves which would double that number in a matter of weeks, this constitutes overwhelmingly the greatest quantity of military firepower ever gathered in a single theatre in the entire course of history. What do these contrary facts mean?

Whatever one's point of view about arms *control*, the future of arms looks problematic. It is assumed that technically stable deterrence has something to do with assured destruction capabilities achieved by both Superpowers during the 1960s. At least we know that, at great cost, these developments provided reassurance not possible with earlier, less stable patterns of competition. Even though assured destruction was sought in greater than 'minimal' terms, the effective technological restriction of both Superpowers' options to large scale, heavily destructive responses, made SALT I an acceptable arms control agreement. But SALT I now looks like an irrelevant arms control agreement, for at the very time it was achieved the Superpowers were beginning to move beyond assured destruction as the basis of their strategic relations. Proliferation of warhead numbers came with MIRV. Accuracy and guidance improvements came with better target surveillance, refined on-board systems, and developed warhead engineering. The implementation of planning for more discriminating use of strategic nuclear weapons came with the capability to rapidly retarget land-based strategic missiles.

Thus proliferating numbers, guidance and accuracy such as to bring a warhead within a few hundred feet of a point target over 6,000 miles, lower yields creating less 'collateral' damage, and command and control systems to make feasible (however improbable) the exchange of only a few warheads: these developments are adding highly discriminating options to the strategic plans of the Superpowers. Yet the consequences of exchanging even a few nuclear weapons remain imponderable. Indeed the special strategic value now accruing to those parts of the Superpowers' forces capable, or putatively capable, of executing limited options, leads to fear of pre-emption. This is nowhere more strongly evident than in the parts of the American 'defence community' which espoused the development of such options in the first place.

Technology, and its human agents, are providing the instruments with which to commit suicide slowly, to promise massive retaliation in small doses, to have an instalment plan for total destruction. This is quite rational if it actually does make deterrence stronger. But how does one know? Nobody is certain. The subject of intra-war deterrence has become dogmatic: we are proceeding by feelings along the route of scientific development. The possibly grave imprudence of such a course can be illustrated by asking what limited-use options will do to crisis dynamics – those items of Superpower behaviour which hitherto have most clearly shown deterrence to be mutual – and by answering (as seems sensible and reasonably predictable) not much, so long as political control and restraint remain high on both sides. But does this suggest a sensible relationship between technical change and politics? Technical change is supposed to *reinforce* stability, to make it less susceptible to shifts of political fortune. But as American fears about Russian show limited strategic options (options advocated by some of the Pentagon's favoured strategists), *are* a kind of war-fighting doctrine; or at least they are bound to be seen as such by the other side. War-fighting doctrines are arguably less rather than more stabilising in relation to *unpredictable*, domestic or international, regime changes. Moreover, intra-war deterrence, a set of concepts prompted by the technology of limited options, inexorably leads to an ugly variant of limited-war thinking, which here includes proponents of theories of 'victory' in strategic nuclear war. Theories of limited war have been shown repeatedly to be fallible at best, misleadingly destructive at worst.

In the European regional situation a different kind of technical procedure, that of alliance politics and bureaucracy, has attended MBFR. The alliances have put into effect force-level and quality improvements such as to make possible rapid mobilisation and intensive standing start offensive and defensive operations. At the same time the conventional military balance has become closely hedged about with battlefield and theatre nuclear weapons. These short- and intermediate-range systems do not possess any sensible structure of deterrence of the kind associated with strategic intercontinental weapons. Especially at the level of longer range systems associated with NATO's doctrine of flexible response, there is a growing uncertainty of purpose and correspondingly greater room for pre-emptive contingency planning. A short sharp war between East and West would probably be such only because it involved early and heavy use of nuclear weapons. In such an eventuality destruction might compare with that of the Second World War over parts of Europe. A long war in the theatre is conceivable only if neither side managed to achieve strategic superiority at the conventional level, and if, somehow, incredible nuclear restraint were observed by all nuclear countries. The result might resemble attrition warfare fought with high technology; it is an awful irony of our civilisation to think of this First World War situation as a preferable outcome.

In the developed world, there is a widening gulf between the *politics* of security and the military relations of the Superpowers and the alliances. In the past, Cold War attitudes, the structure of deterrence, the technological solutions to stabilising Superpower and European arms races made this tolerable. Indeed, it has been cogently argued that an independent logic of military competition *had* to work itself through in the period from Korea to SALT before any meaningful new departure at the level of the politics of security could occur. That argument, whether cast essentially in terms of detente or of arms control, saw a conjunction of the two in the pattern of interests of the great powers. But after many years of anxious waiting, it is clear that the logic of military competition is still working independently of the new departures in political relations that have been achieved. Equally, it is clear that the politics of security have been freed from the logic of military confrontation. We may agree that at present – let alone future – numbers of strategic weapons, the idea of superiority is meaningless; but none the less it is impossible to divorce images of capability from images of intentions. What is evident at that strategic level is also evident, and perhaps more vividly, in the European region, where the precise relationship between particular military balances and international security escapes rational explanation. Here then is a paradox: although the trends of military relations and of other political relations are driving apart, it is upon occasions of crisis, such as Poland recently, that military leaders rely most upon the maintenance of a connection between the two spheres in order to keep deterrence in being, and even invoke detente – otherwise not a favourite term – to that end. Thus, it appears that differing views of the sources of security cannot appeal to military facts alone. These do not speak for themselves; they require explanation. This may be done in several ways.

Perhaps a relationship between arms and politics that makes the best efforts of arms control appear inadequate, actually provides a kind of enduring security in uncertainty which itself controls the *use* of arms, but not their numbers and qualities? This is a critical problem of knowledge. If we suggest, in this manner, that arms control has failed, we are assuming – explicitly or implicitly – a set of connections between technology, international politics and the use of force. For instance, we may say that weapons developments lead foreign policies into institutionalised competition in the developed world, and arms control serves only to maintain an irrational relationship between the qualities and quantities of weapons and their military uses. But such a set of connections is not self evident.

Is it not more modest, and perhaps more accurate, to suggest that perhaps detente and military competition are mutually supportive and reinforcing? This postulates that the logic of technology *provides* the logic of security, and that there is thus no disproportion between improving defences and improving political relations. In these terms we then conclude that arms control is an important artifice for providing the ratification of previously decided arms policies; it is a form of technology guidance. It has not failed, though neither has it achieved its ostensible purposes. It has merely functioned.

Nevertheless, a functional view of arms control and military competition still leaves worrying questions. Is stability sufficiently guranteed by political restraint? Does deterrence depend, after all, upon the rational prospect of disproportionate destruction enduring even through a strategic nuclear exchange, rather than upon the irrational fear of a nuclear threshold which is inelastic in terms either of weapons numbers or indeed of their qualitative attributes? Even if we cannot answer this question, even if it is conceded that a philosophy of deterrence, of the changed place of force in international politics, may be compatible with technology-led politics, another crucial question remains. What ultimate consequences are likely to flow from an emphasis upon technology as the infrastructure of international political relations if, thereby, mutual legitimation of interests is eschewed in favour of ideological competition, notwithstanding a structure of co-operation in other areas?

In the face of these profound doubts, it seems plausible to suggest that arms control has perhaps lost its way, that the drives of technology and ideology and conflict-generating change have proved too strong for a philosophy of reconciliation that does, essentially correctly, view deterrence – at least in its technical assured destruction forms – as ultimately *not* itself capable of bearing the weight of political confidence needed for the operation of a naturally stable international order. It bears this weight *only* because underlying political restraint is greater than the tensions (so far) caused by technology and ideology. But this view posits a thesis which is extraordinary in its implications: that the normal politics of military force has been abnormally sidetracked by a necessary alteration of roles, temporarily. This view carries political choices into unexplored realms, and presents challenges and risks which postwar diplomacy has thus far been unable to face.

In sum, it is possible to argue one of three positions concerning the state of arms control and problems of military power. The first position suggests that arms control has failed. What seems important in this position are two notions: first, arms control processes have not ameliorated Superpower competition in or beyond the developed world; secondly, while the existing relationship between military power, international relations, and the alliances is served by arms control processes, it is an unstable relationship. The second position suggests that arms control has served a purpose, rather than succeeding or failing. At best it has stabilised an unavoidable Superpower competition; at worst it has occluded that competition. In the end, however, the Evolved Cold War has incorporated, but not depended upon, arms control. The third position suggests that arms control has failed to achieve objectives independent of managed strategic competition; as a dependent variable of managed competition in the Evolved Cold War it has served as a palliative and thereby disappointed its more ambitious proponents and most philistine critics; but stability is no more automatic than instability and, both depend upon perceptions of change and more or less prudent responses.

The three positions examined above are schematic of underlying assumptions and hypotheses from which other permutations can be derived. All three positions entail problems of knowledge about the relationship between military power and arms control. Among them all is a common factor: an acceptance that in greater or lesser degree the relationship can be managed, and that the process of management is susceptible to political influence. But it is impossible here to be decisive about one position or another, and whilst it is perhaps tempting to enlarge upon whichever explanation appears to maximise the potential effects of political will upon change, neither adjudication nor preference provides a logic of development forward from this point. Rather it appears necessary to rely upon inference, and in so doing to be cautious about statements sporting certainty. The major inferences here are that the criteria of change in the present state of arms control and military power depend entirely upon the presence or absence of *political will* to affect, first, the nature of Superpower competition and, secondly, the structural relationship of security between developed and developing worlds. But arms control is in a deep dilemma, caught between perceptions of long-run change which appear to destabilise developed world international relations, and dependent upon political and military relationships which have entrammelled arms control processes in serving short-run managed competition. It is to some elements of this problem that the balance of this chapter is addressed.

Military Science as Transnational Ideology

According to a children's television series popular in Britain for some years, the Daleks are machine-creatures which had their origin in the technological fanaticism of a demented, vengeful scientist, himself half transformed into machine. The Daleks are hopelessly incomplete by comparison with the human beings they periodically oppress and constantly plan to subjugate or destroy. The principle they operate upon is that of obedience, which they impose and obey absolutely. Daleks cannot tolerate disobedient Earth people. But neither do they possess the faculties to apprehend humanity in its full non-logical complexity. Daleks are powerful. By means of superior technology, rigid organisation and absolute obedience, they can exterminate and dominate: they are the Roman Legions of Space. They cannot understand, let alone question, the limitations of power gained from technology, nor the objectless nature of expansion for the sake of itself. Imbued with the will to conquer, to change, to develop what they dominate in the rigid mould of allegiance to scientific social development, they always – inevitably as it seems – lose out to Serendipity, happily symbolised by an altruistic genius, Dr Who, whose fortuitous arrival through time-warps coincides with periods of Man's greatest peril.

In the contemporary relationship between weapons development, alliance politics and the management of deterrence, there is something, incipiently at least, like the struggle between the half-human traits of the technological Daleks, and the half-materialist existence of humankind. We have set technology to serve political objects by means of military power, but have forgotten that the problem of power in politics is manageable ultimately only by an understanding of interests and will, rather than by surrogate organisations that impose technically and doctrinally narrow attitudes upon us.

None of the commonly held objectives of present day culture are sustainable in a world of states without collective social constraints. Systematic constraints must include constraints upon the exercise of military power. The deterrent qualities of nuclear weapons, the collective self-defence provisions of alliances, the limitations of crisis behaviour through signalling and communications, the monitoring of agreements and movements through surveillance and national technical means of verification, the ever-improving command and control of military forces in being: all are features of a systematic institutionalisation, more important in themselves than new weapons technologies. The control of technology is therefore undeniably crucial to the kinds of constraints which have been built into the international political relations of the developed world. But the substitution of technological means of constraint for real adjustments of interest and political will, is establishing strong features of Dalek-like attitudes on both sides of the East-West divide. These features follow upon, and reciprocate, pre-existing ideological antipathies, themselves redolent of political interests suspended within deterministic ideas of national self-development and destiny.

On the Soviet side this has taken the form of an unnatural development of military capabilities among the activities of the state. The Soviet Union has become disfigured, like a small man with disproportionately long and muscular arms. This overdevelopment of its military capabilities is attributable to its political will to defend its territorial integrity, to project its status as a great power, and to break the frustrating bonds of Containment wherever they have been established. In turn, however, that persistent will has been founded upon an unreal appreciation of the dangers facing the Soviet Union from capitalist states, and upon a bloated estimation of the leading role of the Soviet Union as a bastion, and the only legitimate authority of Socialism, either in One Country, or in the Socialist Commonwealth, or beyond. Its failure to establish (or for that matter to choose) a European identity; its tardy industrialisation; its Civil War following the Revolution; its rejection by Britain, France, and smaller European states during a period of profound peril from Nazi imperialism; the unimaginable political revulsion stirred by over twenty million deaths in four years against the Axis; the insecurity which followed upon galvanising the far mightier power of the United States against it; the profound strategic risks which must have periodically appeared in assessments of the nuclear balance between the Superpowers: all of these historical experiences, most within living memory of the present generation of leaders, all deeply significant to the Russian view of international relations as well as to the Communist view of Marxist-Leninist praxis, have served to distort the Soviet view of reality, to inhibit its appreciation of the independent well-springs of contemporary change, and to reinforce its antiquarian set of beliefs about Soviet socialism as a force for World Revolution.

It is also essential to appreciate the extent to which science and its organisation have combined to form ideological elements in American foreign policy. The historical basis of this is more straightforwardly economic than in the case of Russia, for the United States participated in, and soon came to lead, the second and subsequent waves of industrialisation that caught up Europe and what has become coextensive with the developed world in the late nineteenth century. It was perhaps the absence of any security imperative in the growth of American power that made it so innocently enthusiastic for the leading role of science and technology. After all if there is no major threat to a society expanding economically it will tend to expand its power economically, and provide rationales for this in terms of its national security interests. It will also imagine perfectibility in defence to be a matter of judiciously intertwining organisation and science. Consider the exercise of American power in Latin America, the Caribbean, the Pacific and China, Mahan's image of sea power as the ideal form of military strategy in a world economic context, and, the equally misplaced idealisation of air power which lay deep within postwar American thinking about Containment and retaliation against Soviet expansion.

Three principal strategic changes account for the difference between Mahan's easy assurance of the merits of a blue water navy and Dulles's latent defensiveness in theatening Massive Retaliation. The experience of two world wars demonstrated the human carnage which accompanies scientific development in war. Secondly, the emergence of the Soviet Union as a challenge to American global interests, and doughty opponent of conceptions of harmonious world capitalism, focused American strategic thought upon specific analyses of threats, interests and commitments, and obliged the operation of American economic power for the first time to follow, rather than lead, a definition of military balance. Thirdly, nuclear weapons, at least after the Soviet Union demonstrated its own, transformed the fact of considerable geostrategic immunity from attack, into the fact of considerable and dramatically growing vulnerability to immense destruction.

The social psychological response to this sudden inversion of the terms of national security was fearful, as was evidenced both in McCarthyism and in the alleged – but mistaken – appearance of a 'bomber gap', and a 'missile gap', favouring the Soviet Union. In circumstances of objectively considerable change, and subjectively considerable fear, the United States viewed its policy as dependent upon well-organised national leadership utilising to the utmost the power of science to produce technological solutions for its defence, and the defence of its worldwide interests and commitments. Kennedy's Great Leap forward; the sad escalation of the Vietnam war far beyond meaningful ends-means assessment; the continuous conflation of estimates of Soviet capabilities so that what capabilities the United States needs always exceeds previous definition, and tends to conform with future capacity to produce and pay; and the constant inducement to arrive at 'higher-level' theoretical explanations of strategic deterrence: these and other examples provide evidence of a strongly entrenched, bureaucratically organised, science-dependent ideology of national security. The subjective adjuncts of this ideology are Soviet expansionism, Western defensiveness and weakness, and the crucial reliance upon permanent, science-based, military preparedness and weapons development.

The defence of Western Europe and the highly formalised rhetoric of alliance relations provides another, independent yet reinforcing, source of Dalek-like thought. After the Second World War the countries of Western Europe were too weak militarily to have prevented the Soviet Union from asserting a European political identity, in the end in terms of its own choosing. An East-West divide was precipitated. It was based upon fear of Soviet intentions in Europe, particularly towards Germany, the prerequisites of economic recovery in the capitalist international economy, the historical, cultural, and political shared values which gave North America and Western Europe common cause to ally together to define the geopolitical terms of collective defence, and the ability of technology-based American military power to offset NATO's conventional military weakness. With the development of strategic nuclear competition between the Superpowers, the Europeans came to depend upon the exercise of a particular form of American power: the perceived ability to deter aggression by the threat to share the destructiveness of nuclear war in the defence of shared values. West Europeans have since clung to ideas of extended deterrence with a tenacity which the United States has increasingly found uncomfortable, but at the same time the Europeans have refused either to accept monopolistic American strategic control or to relinquish their dependence upon American conventional forces. They have been enabled to do this in part because the availability of American technology in an alliance context has posed real strategic problems which the United States could not successfully have managed without co-operation. In effect, the United States has been prepared to pay considerable ground rent for basing its forces forward in Europe, and the Europeans have implicitly chosen to take payment in the form of strategic reassurance and intra-alliance co-operation. However, this quite efficacious bargain, or set of bargains, has required a particularly strong ideological link between leadership, power and military technology, especially since NATO provided itself with a nominally unanimous strategic doctrine, Flexible Response, in the mid-1960s.

The problem which now emerges, and explains the choice of the Dalek metaphor, is not simply that the United States, or the Soviet Union, or Western Europe, or – beyond my main focus – the newly independent and rapidly developing military powers of the developing world, has a present, or incipient, ideology of power which is profoundly and mistakenly over-balanced in its belief in the efficacy of scientific weapons development. Rather it is the case that each of the three main cultural regions views the *other two* as susceptible to this fallacious belief, ascribing its source to the unnatural materialism of the underlying political ideology. The United States views the Soviet Union as materially backward and ideologically militaristic, and the Soviet Union views the United States as dangerously overdeveloped materially and militarily due to monopoly capitalism, and Western Europe as the raddled handmaiden of this international structure of domination. Western Europe views the United States as overdependent upon mechanistic, technical approaches to political problems, and the Soviet Union as doggedly, hopelessly, committed to a positivistic creed of social development that extinguishes freedom of thought and engenders conformism.

It is not suggested here that each side is equally correct, only that each is in different ways affected by a comparable phenomenon. Neither is it possible here to disentangle the social and intellectual roots of these problems, which are intractably caught up with enduring philosophical questions about Man and continuously developing states of economic transformation. These deeper problems *do* affect international politics,despite conventions and orthodoxies of thought which see international relations as fundamentally different to those of social relations within the domestic community. The way in which their effect is depicted here is as that of a carapace of perceptions wherein each point of strategic view sees clearly the dangers and disfunctions of Dalek-like thoughts and policies in the strategic perspectives of the others; but this carapace is a property of the international structure of military power and deterrence within which *all* – Americans, Russians, Europeans, the alliances – are participants.

The structure of the system which is generating this behaviour is not deterministic; it is a holistic outcome of the interactions between differing strategic positions. But non-participation, unilateral disarmament, relinquishing independent deterrents, or leaving NATO does not solve the problem. The problem is that between the wayward, naturalness of our human condition which we would wish to see expressed in detente and arms control, and the puritanical determinism of dogmatic authority which denies the possibility of fundamental compromise to anyone while insisting upon and expecting it from all others, there stands power, leadership and the catalysis of science. Power is presently inalienable from international relations. Leadership is in any case necessary and science is ethically neutral and in itself cannot provide the terms of avoiding a surrender of human qualities to automaton forms. It is, then, to the way in which we *think about* power and war, to how our attitudes shape the pusuit of security through technology and frustrate aspirations for arms control, that we must turn.

Questions of Capabilities and Intentions

An example of the transnational structure of the relationship between ideologies and technology can be provided in the debate about Soviet defence spending and intentions, and in contrasting the typical attitudes found in this debate with 'domestic' explanations of American strategic weapons development.

In order to depict explanations of the motives and objectives of the other side, I choose to quote at length from a recent analysis of Soviet strategic behaviour by an American expert on this subject. Because the following passages reflect an expert view of Soviet strategy which is not criticised as such, it is important to note that the passages cited are from the introduction to the author's substantive discussion, and that the ideas of interest here are those of the United States' *self image*. The author begins by pointing to certain, allegedly ethnocentric strategic assumptions common in the 1960s' atmosphere of detente and arms control – that the Superpowers would gradually evolve a common strategic view of stability and deterrence, based largely upon objective truths, uncovered by American analysts in the postwar debates over the implications of nuclear weapons. He continues:

> In the light of the repeated frustrations encountered throughout SALT II and the growing evidence of Soviet determination to acquire forces well beyond those required to support a simple 'assured destruction' policy, however, it has now become clear to all but the most hopeless romantics that the time for such optimism has long since passed and that a considerably more sober appraisal of Soviet motivations and goals is in order. From every indication that one can gather from its uninterrupted record of force enhancement since the conclusion of SALT I seven years ago, the Soviet leader-

> ship has signalled its unambiguous commitment to the accretion of a credible war-fighting capability in all bands of the conflict spectrum in total disregard of, if not outright contempt for, repeatedly articulated Western security sensitivities. Although this commitment does not mean that the Soviet Union is any less interested than her Western counterparts in the continued avoidance of nuclear war, it does suggest an underlying Soviet conception of deterrence quite unlike that which had traditionally held sway in the United States. Accordingly, the American defence community has found itself increasingly driven to base its future planning on the discomfiting reality of demonstrated Soviet performance rather than on the evanescent hope of the eventual convergence of Soviet and American strategic values . . .[2]
>
> To begin with, the Soviet Union defines the nuclear dilemma and the force requirements she sees as dictated by it using an intellectual approach quite alien to the concepts that have largely informed the strategic policies of the United States over the last two decades. To put the point of critical difference in a nutshell, the American propensity – running as far back as the formative works of Bernard Brodie in the late 1940s – has been to regard nuclear weapons as fundamentally different from all other forms of military firepower. Naturally, flowing from this appreciation has been a consuming American belief that any widespread employment of these weapons would call down such unmitigated calamity on all participants as to make a mockery of the traditional Clausewitzian portrayal of war as a purposeful tool of national policy.[3]

The author proceeds to contend that as a direct consequence of these strategic beliefs, the United States rejected strategic superiority as a force posture goal, abandoned efforts to provide for significant defence against nuclear attack, refused to seek comprehensive hard-target capabilities that might threaten the survivability of Soviet strategic retaliatory forces, and sought through SALT to ensure that mutual vulnerability remained a mechanism for regulating stability. Conversely, the author contends that the Soviet Union has refused to accept any stabilising mechanism such as that of Mutual Assured Destruction, has continued to view nuclear war as 'far from inconceivable', and has sought to retain war-fighting capabilities with weapons, and methods of active and passive defence against nuclear attack.

What is wrong with the self-image implied so strongly here, and with the criteria which it generates for judging Soviet capabilities? Fundamentally, it is a view which utterly failes to note any dissonance between what American strategists have propounded as the most reasonable beliefs upon which to found a secure nuclear posture for the United States, and what American leaders have *in fact* decided to be the appropriate force posture. It is a view which takes an indulgent and over-conservative view of American capabilities (and therefore, in the author's logic, intentions also). Moreover, rather than attempting to provide independent criteria for international stability or Soviet strategic purposes – in other words a distinct context within which to attempt to make sense of Soviet capabilities *and* Soviet intentions in relation to one another – the author assumes the political innocence and universality of weapons technology as an objective measure of strategy, and reasons that because in these terms American strategy can be seen to be 'good', and Soviet strategy can be seen to be 'not good', therefore *American* strategy proves that *Soviet* strategy is antithetical to goodness. This fails to differentiate between what the United States may individually define as 'good' for its own national ends, and what is objectively 'good' for international strategic stability. These criticisms are worth expanding briefly.

The United States may be said to have rejected strategic superiority as a force posture goal if one looks at the logic of deterrence distilled by the foremost civilian strategists, or if one looks at the numerical balance of strategic nuclear launch vehicles between the Superpowers over the last twenty years. As regards the former – and leaving aside the ambiguities presented by extended deterrence, by damage limitation and by escalation control, aspects of deterrence which have always influenced American force posture planning and intrigued American strategic thinkers – it is reasonable to suggest that Bernard Brodie represented what was to become the mainstream when he contended that an intuitive map of deterrent effect could be plotted with numbers of weapons graphed against 'deterrent effect', the result looking like a very steep curve rising to a more gradually attained plateau.[4] The steepest part of the curve represented the deterrent effect deriving from the certain delivery to targets of only a relatively small number of weapons, Brodie suggested, targeted upon major Soviet cities. The plateau would be attainable only with strategic superiority, presumed to mean dominance in all phases of nuclear exchange, and would be attainable only by massive numbers of deliverable warheads. Brodie went on to suggest that prompt action should be taken to head off technological proliferation of weapons systems which would complicate management and perception of stability in these terms.

Elsewhere Brodie argued that controlling escalation is a chimera because the dynamics of violence and the dynamics of deterrence in war would operate against each other in such a way as to make rational, deliberate, escalation *very hard to do*, and unintended, irrational escalation *very easy to occur.*[5] These and similar beliefs, based on much rigorous analysis, have informed the policies and doctrines of successive Secretaries of Defense and the Administrations they worked for.

Yet notwithstanding all of this, while the United States has sought stability through arms control, and in those negotiations has sought assiduously to persuade the Soviet Union of the merits of the mutual assured destruction relationship, it has also pursued qualitative and quantitative weapons improvements in such a way as to render meaningless the contention – based only on a count of launcher numbers – that it has relinquished superiority as a force posture goal. If one compares the number of strategic nuclear delivery vehicles with that of launch vehicles, one finds that while the latter has remained constant since long before the signing of SALT I in 1972, the former has risen several fold to between 9,000 and 11,000. Much more importantly, the accuracy of American delivery vehicles has increased by at least 50 per cent for most of its strategic forces during the past decade, and for certain parts of its force the degree of improvement has been far greater. As is commonly acknowledged, a halving of Circular Error Probable (CEP) allows a tenfold reduction in yield to give the same 'kill' probability against a given target. Although American warhead yields have tended to be reduced – smaller warheads mean more warheads per launch vehicle, or more decoys, and these smaller warheads are less easily targeted by missile defences – the military utility and 'lethality' of American strategic forces has increased greatly since SALT I.

Moreover the particular combination of guidance, accuracy, weight and number improvements undertaken by the United States is useful only in targeting strategies for hardened, point, or relatively small area targets such as are typically represented by military forces and infrastructures, including missile silos, command bunkers, radars, airfields, railway marshalling yards, specific industrial locations, and so forth. In short, the United States has pursued, is pursuing, and will continue to pursue even within the terms negotiated for SALT II, a strategic doctrine of 'Equivalent Countervalue' rather than Assured Destruction. Equivalent Countervalue means essentially the displacement of the important distinction made in early deterrence theory between counterforce and countervalue, by regarding the enemy's military capabilities (*including* his military-industrial complex) as the appropriate focus of the bulk of strategic forces.

It is not necessary to explain an Equivalent Countervalue strategy as *intentionally* opposed to an Assured Destruction strategy. One can, for instance, look at the practical difficulties of assessing what is 'enough' in Assured Destruction terms. The following passages illustrate this point, treating the relationship between capabilities and intentions from a 'domestic' American point of view.

> A strategic planner . . . must simultaneously deal with (1) 'sizing' the strategic forces of the 1990s via decisions now about development and production; and (2) with 'employing' the strategic forces that he has inherited from the development and production decisions that his predecessors made, roughly speaking, a decade or more ago. Currently the 'sizing' decisions are, and have been since the mid-1960s determined mainly by the criterion of Mutual Assured Destruction (MAD). But the numerical tests to implement that criterion are so devised that, *if* the tests are applied properly, you will still buy more strategic forces than will be needed for a MAD operational capability a decade hence. The *sizing* decisions apply the *philosophy* of modern statistical decision-making, under conditions of uncertainty. In McNamara's words: 'When calculating the force required, we must be conservative in *all* our estimates of both a potential aggressor's capabilities and his intentions. Security depends upon assuming a *worst* plausible case, and having the capability to cope with it.'[6]

The author proceeds to point out that risk and uncertainty require quite different approaches. Whereas the former is susceptible of fairly rigorous analysis based upon assumptions of rationality, the latter is in essence a form of rationalistic gambling, not different in nature from Pascal's famous conundrum, but with a malevolent enemy filling here the equivalent place of a putatively beneficient God: one is not betting so much that the enemy is malevolent, or even that in ten years he will be an enemy; one is betting *against* one's own scientific rationality.

> In sum . . . if we use the proper statistician's test for *high confidence* in procurement decisions now, even if tested by a 'Mutual Assured Destruction' (MAD) criterion, we will buy a force for use a decade later that will be ample for flexible use in a variety of situations, in

> appropriately different ways, as the Commander in Chief may choose to use it. That force should also have qualitative attributes for flexible employment. If, however, you use a *low* confidence criterion for purchasing a 'MAD' capability, our President in the 1990s will find that his force is so small that it cannot be flexibly used. He will be in deep trouble . . .[7]

The questions which arise from this kind of exposition of the operational analysis of MAD, and the changing criteria of a deterrence posture that it not only reacts to, but evidently also *creates* a need for, are as follows. Is there any meaningful distinction between a deterrence posture based upon MAD, and one based upon warfighting? If the most meaningful distinction is at the level of *intentions*, how is it possible to devise empirical tests for these which will incorporate adequate respect for the statistical logic evident in sizing our own strategic forces on a MAD basis, but with worst plausible future cases in mind? In arms control negotiations, has the United States ever attempted to persuade the Soviet Union to agree to a minimal deterrence definition of MAD, and if so was it prepared to forego the flexible options that were foreseen as part of its own posture? If in fact a very similar form of the statistical logic of risk plus uncertainty based upon fearful beliefs is available to the Russians, can it legitimately be alleged that they are pursuing force goals incompatible with Western security interests, or must it be acknowledged that, within a different context and within possibly different images of themselves and the world, the force postures of both Superpowers are evolving approximately similar capabilities because of approximately similar security interests? If this, or something like it, is the case, then how can national leaderships be made to acknowledge explicitly that the strategic policies of both Superpowers are set within a shared context of technological development, itself dictated by an apparently scientific form of rational analysis, the roots of which in (probably) both cases are fear of uncertainty, an image of malevolence, and a perceived refuge in technology? These questions cannot be pursued further here. The rationality of using and threatening force even in an age of overwhelming power is best approached through Clausewitz.

The Relevance and Ambiguities of Clausewitz

Clausewitz concerned himself mainly with the inner dynamics of war.

In so doing he drew heavily upon historical evidence and direct experience. But he sought to free his analysis from tight temporal-spatial references. Thereby he succeeded in identifying some of the most enduring characteristics of modern war: its innate violence and transcendence of reason; its tendency to concentrate and escalate using all available means; its political logic, which shapes war as a social phenomenon and provides it with a rationale; those manifold aspects of 'friction' which in reality modify war's theoretical absoluteness and return it to a strategic calculus of ends and means, dominated by poltical interests; and the dualistic character of real war, where political logic and military aims intertwine continuously, transforming an analytically clear means-ends relationship into a true dialectic of opposing principles, one based upon the rationality of social values, the other upon the passionate pursuit of victory. In all of this and because of its admixture, Clausewitz argued less emphatically but no less importantly that there must be a simultaneity of responsibility for both the conduct of war and for the conditions of peace. Far from either disjoining responsibility for the conduct of war from civil government, or fusing civil and military authority together, this simultaneous responsibility can be carried through only by a division of functions arbitrated by purposive civilian authority.

Inevitably the focus of 'On War' entailed opportunity costs. There is no treatment of the social forces which motivate the governments of status to use the instrumentality of war. There is scant and always elliptic reference to the structure of international politics as a context of states' actions. There is not much consideration of material conditions of warfare, such as science and technology, operating upon the relationship of state and society autochthonously. Still, the political logic of sovereignty that provides the modern morality of war; the relativity and fungibility of power that provides the basis of the means-ends calculus of political interests and costs; the organic nature of the societal motives of security that result in forces, weapons, organisations, for war: all of this can be readily interpolated into and around what Clausewitz so elegantly wrote.

Here then is a commodious framework for approaching the place of strategic factors in international relations, and the manifold changes which have occurred in the study and practice of strategy since. Strategy becomes the attempted imposition of a rational plan of action upon recalcitrant reality. In interpersonal or social relations it is accompanied by a variable and finite expectancy of conflict, in aims or over objects. In interstate relations it builds this expectancy of conflict into an expectancy of violence. Although sovereignty legitimises the uses of force to effect changes in the structure of international relations, the inefficient realisation of that power in war means that change by force occurs in a highly resistive medium. Hence war both represents a continuation of politics as a strategy of violence, and a special state of resistance in which outcomes are not logically deducible from means and ends formulated analytically. Strategy *in* war, then, becomes the use of mobilised force to violently compel an opponent to submit his will; and the strategy *of* war becomes the calculus of securing objectives in terms of interests where, in the normal context of international relations to which the end of war returns the states, security is never in itself an attainable objective. Clausewitz clearly indicates that war, in itself, settles nothing finally.

The dominant motive of security and the incalculable violence of war meet together in a way which paradoxically distinguishes power and force. Power is the currency into which states must exchange the domestically formulated interests and values of their policies. Power is reflexive, a necessary response to the resistive – indeed compulsive – environment of international relations. Its theoretical foundation is sovereignty, whether defined by traditional legal-political theory or judged as an expression of behavioural autonomy, but its structure is that of interdependence of objectives in a notionally independent state.

Force on the other hand is elective, a continuation by other means of attempts to effect short-run change in the resistive structure of international relations. Its theoretical foundation is opposition of interests in greater or lesser degree. Yet in practice it is sovereignty, whether legal or behavioural, which provides the basis for conducting war. Power offers potential and influence, but only a negative measurement of action. Force is something positive, but in aiming at positive outcomes it carries the cost of changing other conditions – in the domestic environment through the inefficient mobilisation of power and the social consequences of war upon society, and in the international environment by altering the terms of interdependence, or even the wider structure of relations between powers not directly involved in war. The domestic and international effects of the Vietnam and Middle East Wars are but examples.

The interacting process of war which results from this bringing together of social structures, powers, forces and wills, is so unpredictably dynamic that it is possible at best to discern a relationship between escalation and declension on the one hand, and strong national control by civil authority, on the other. It is then possible to suggest that the intra-war dynamic of reason and passion on both sides is likely to be more formative of the actual character of any given war, than are the pre-war, unengaged, assessments of interactive means-ends relations calculated analytically. It is also possible to suggest that the nature of the dynamic in war will be crucially affected by the strength of the interests and underlying political will present at the outset. It is not possible to go beyond these generalisations to formulate a theory of war that would provide an explanation in terms of cause, or a logic of development between the social antecedents of war and the interplay of social forces in war. Nevertheless, Clausewitz, and what can be read around his work, provides us with some important messages for today.

War is inherently unreasonable. It is an activity of violence unbounded by any imperative other than victory, the triumph of will through sheer force. Many modern interpreters of Clausewitz would quibble with such a judgement as made here, for indeed Clausewitz is mainly interested in violence directed by will rather than passion. There is a tendency to depict even his Absolute war as a dialectic of opposing wills, as a violent interaction dominated by a kind of 'rationality of the irrational'; but this is to miss the essential point. Clausewitz allows that in modern times 'hostile intent' may be ruled by reason. However, he then proceeds to demonstrate that the presence of such reason *does not* inherently limit violence; the mutual reasoning of eachparty interacts in such a way as to cause escalation to the utmost concentration of effort. Today then, as we play with our rational formulations of deterrence, it is worth remembering that Clausewitz has shown this has *no* consequent moderating effect upon a violent conflict in which each side is attempting to overpower the other.

What modifies warfare in reality is not reason itself, still less analytical models of cost-benefit strategies, but rather a social and political structure of government operating in a context of international relations and pursuing political objectives by means of war that do not amount to, and can therefore rationally be distinguished from, total victory. Put negatively, what does *not* moderate warfare in reality is absence of cost as the limiting case of social and political will.

Today we see the next stage of weapons development being urged in part in terms of a 'theory of victory' as a complementary rationale to the concept of intra-war deterrence, both being concepts used to justify the growth of war-fighting capabilities. Clausewitz is not unambiguous here, but he does seem to indicate clearly that a 'theory of victory' cannot claim to be a limiting factor in war.

It is not, then, *utilitarian* costs that limit will, as the United States found in Vietnam. Economic costs can be offset directly *by* will, as when a country undertakes greater sacrifices to sustain a war, for independence, or against an invader, or for an ideology, or for some combination of all three. Value resides in objectives no more, in this sense, than it does in labour in Marxist economics. Value resides in the transvaluation of individual choice. Where, then, are the social costs which are to be the limiting case of social will in the pursuit of aims in war? It is what Clausewitz meant most generally by the 'centre of gravity' of the enemy. It is wherever the social organisation of the enemy's will to pursue interests *and* endure costs resides and comes together. As a young man I remember American prognoses of North Vietnam's will to continue bearing the costs of the war of reunification being based on estimates of how much of their younger generation had died, on top of how much of their older generation was already dead. I also remember a satire upon this absurd positivism by an acute satirical publication which showed a long line of bodies laid out as in evidence, with a caption which said something like, '. . . 1001, 1002, 1003, . . . and they must all be 'Cong, because they are dead'.

The centre of gravity of a society may be its armed forces, its territory, its economy, its populace, its industry, its capital, but what decides which of these, or which combination, or which other, is social structure, primarily the structure of concepts of government, authority, and legitimacy. In other words, it is the nature of community, its values, and the relationship between those who fight and die, and those who organise and direct. It was for this reason, it seems, that Clausewitz insisted upon the profound distinction between the competence of a government to decide political war aims, and the *lack* of competence of the military to do this. Legitimate civil authority alone is capable of estimating the social costs which are to be the limiting case of will in the use of organised violence for political ends in international affairs. What relationship should society insist upon between the scientific military planning for war undertaken today, and the costs which would be tolerated in the defence or pursuit of any of the apparently endless lists of putatively limited options?

Post-war American (or British and French) civilian strategists have not seen Clausewitz as outmoded by changes in the use of force brought about by nuclear weapons. The reverse is the case: most of the principal civilian strategists have been influenced by Clausewitz's thought, and among that number are some who see arms control as a central part of the structure of a stable deterrent relationship between the Superpowers. The significance of this reveals further ambiguities in the notion of rationality implied by Clausewitz and accepted – but perhaps misconstrued – today.

Rationality, Community and War

When Clausewitz says that in order for war to be rational it must be fought for political objectives decided by legitimate civilian authority, and that military men – and by extension, scientists and technologists – have no business deciding these objectives, what he means essentially is that there is an organic sense of rationality which inheres in the relationship between a state and its society. When Clausewitz says that the military man has no business to formulate plans for war without first requiring instructions as to the political object which provide the purpose of war, he is saying that there is another kind of rationality, namely the rationality of military aim, in which there indeed is no substitute for victory.

The first kind of rationality is what might be termed constitutive rationality. This is rationality which *cannot* be replicated in externally observable terms, though there are indeed external criteria which can be used to judge it. The rule of non-contradiction appears to be applicable to the relationship between belief and action. The notion of 'internal' causality is another criterion: the presumption that actions are predicated upon beliefs. It is possible to invoke a criterion of 'sensibility', that action is proportioned to attainable forms of fundamental beliefs. Constitutive rationality, which provides appropriate terms of explanation for individual and social action, including political action in the name of the state, is, then, determined not by external rules, but by the internalisation of principles of *consistency* between belief and action, the basis of which appears to be values as beliefs.

The second kind of rationality is *demonstrative* rationality. This is rationality which can *only* be judged in externally observable terms. It is the kind of rationality which has been adopted methodologically in most scientific and aspiring scientific disciplines where *empirically testable hypotheses* form the *modus operandi* of decision making. Demonstrative rationality in social science requires the notion of ascertainable value, or as it is more commonly referred to, utility. Utility, at least in its non-Marxist forms, does not necessarily presume the world of matter to be the source of all values. But it does presume at least that all material entities are valuable; they carry assigned values. In conflict over objects, therefore, the amount of effort which it is rational for one to make to gain any objective cannot exceed the value accruing through its gain. As it is very difficult to apply this test under the dynamic conditions of war, where in any given fight the amount of effort *actually* made is going to be as much as function of the effort made by the enemy (whose valuation of the object in question may be different, and who is also being influenced by my effort) as of the value assigned to the object, there is need of some objective reference condition for the setting of military aims. This reference condition, perhaps unavoidably, is victory, the military equivalent of the triumph of will. Of course victory may not correspond with the political purposes of the war. Hence there will often be great tension between the rationality of military aim and the rationality of political purpose. Moreover the source of this tension seems to be a categorical difference in modes of rational evaluation.

But what about the situation where the object of the war is *itself* to limit its proportions more strictly than to what is not constitutively irrational? Clausewitz is clear on the principle of dividing responsibility for evaluating political objects from that for the management of military aims, and about the relative position of each in the conduct of war. He is not clear as to whether there is only one category of war – Absolute war – from which all real wars are derogations of degree; or whether there is a second category of war – Limited war – within which fall all those wars, distinguishable from other real wars which tend to escalate, by the presence of particular rational constraints that eschew the fullest possible use of available means. This ambiguity matters profoundly when we move from the interpretation of Clausewitz back to the relevance of his analysis today, because it illumines the relationship between rationality and the definition of political purpose. But in order to come back to this in the context of arms control, it is helpful to diverge somewhat, to consider what has happened, in historical terms, to that organic community of state-and-society-in-international-relations, which helped make it so simple for Clausewitz to distinguish legitimate from non-legitimate authority in the setting of war aims.

The period from 1815-54 was one of relative stability in great power relations. There were no wars between the major powers of Europe, and few wars between the greats and others. Notwithstanding this relatively pacific external picture of war, domestic societies were undergoing profound social change, which promised revolution in most countries and produced it in some. In Western Europe, these revolutionary movements were intimately associated with the process of industrialisation, though in other areas such as Bohemia and Hungary, proto-nationalisms developed that were more akin to France before 1789. The contrast between external peace and balance, and internal unrest, is instructive. External balance was due to many factors: the nature of the Vienna Settlement which incorporated the defeated as well as victors in the terms of peacemaking; the Concert principle of collective responsibility for the international stability of Europe; its institutionalisation was able to emerge as a flexible order of intermingled Treaty law and rules of interest; a conservative *status quo* which transcended ideological divisions between Absolutism and Constitutionalism; a balance of power geographically described by militarily strong peripheral states whose principal capabilities were asymmetrical (naval and land forces), and a militarily weak centre; reasonably clear spheres of interest; and indeed, in most cases, strong internal preoccupations. Withal it appears crucial that the public order of Europe was an idea which, however querulously interpreted from time to time, was shared in common by statesmen and elites whose cultural understanding of each other formed an international community of interest, stronger in structure and substance than the putative national communities founded by revolutionary interprepations of sovereignty. In this sense international relations displayed an organic sense of community and a constitutive sense of rationality. Recognition was unreserved; interests were mutually legitimated. Deep divisions of belief and antagonisms of policy existed, but they were subsidiary to international order.

From 1854 the Vienna settlement was increasingly threatened by disharmony between the great powers. Although the Crimean War was a limited war in classic terms, it was also, however fumbling and accidental its origins, a war that in its unintended consequences did much to undo the community of interest which had bound the powers together. Russia was alienated; an inherently conservative power became effectively revisionist because of resentment against the Black Sea clauses of the Treaty of Paris. At the same time this permitted a critical increase in the room for manoeuvre of three more revisionist powers; France under Louis Napoleon; Prussia, whose foreign policy was dominated by Bismarck from 1860; and Piedmont-Sardinia. French revisionism was always the least clear sighted and most precarious, for France – as opposed to its leaders – had little to gain by destabilising the *status quo* that would not have accrued with time. Prussia, by contrast, could seek to translate its growing economic power in North Germany into some readjustment of political power there, where Austria continued to assert its historic primacy. Piedmont-Sardinia could use the sentiment generally favouring Italian Unification, but this was most effective when in turn it was used by one of the other two overtly revisionist powers as a means of exerting pressure against the structure of the system, as France used it in the war of 1859, and Prussia in the war of 1866, to drive Austria to distraction. The Habsburg Empire was itself internally unbalanced and thus susceptible to responding aggressively to externally pointed threats, a strategy used disastrously by France in its interventions in Italian and South German parts of the Empire, and successfully by Bismarck in his wars against Denmark, Austria itself, and France. Paris in 1856 was not the last Congress, but it was the last Congress assembled to ratify the outcome of a war between the great powers, before 1918. None the less the wars between 1854 and 1870 continued to be wars of revisionism within what was recognisably the post-1815 settlement.

What constituted a change of system in 1870? It was not so much a change in the balance of power, though henceforth it would be dominated geographically by a powerful German state, as a change in the conduct of war. This in turn exemplified and accelerated change in the nature of state-and-society-in-international-relations. The period on either side of the Franco-Prussian War made for a transition to a different order, with different concepts of society, domestic and international.

Between 1854 and 1870 there occurred a revolution in warfare. This has been so lucidly written about by Michael Howard[8] that there is no need to document it further. After the war ended, the revolution continued. The application of new technologies and scientific techniques, themselves societies and political structures, began to alter profoundly the sources of conduct of states in international relations and hence the structure of relations between them.

Diplomacy could be conducted primarily on the basis of a concert of interests binding elites by common culture, history and tradition. It was this fundamental fact which made Gladstone wrong and Bismarck right. Moreover, military power could not continue to be exercised on the basis of forces in being. Foreign policy had been an activity divided off from domestic politics. The advent of mass warfare in conditions of mass politics, both being transformed and brought into a new relationship with one another by economic growth and industrialisation, meant the adoption by the state of *internally* legitimated, and increasingly also *internally* generated, notions of national interest.

International security came to depend crucially upon the state's mobilisation base, its national industries, its technology, science and population. In turn, the international behaviour of states was shaped less and less by the perceived limits imposed by a communally recognised European order, and more and more by the potential capabilities provided by a nationally ordered set of interests. This is not to say that state-society relations were brought into the field of foreign policy for the first time, for national interest was a long-established precept in great power relations. Nor is it to say that states ceased to be motivated by the consideration of international society, for much of the economic and cultural relations of Europe and the wider world continued to be conducted without reference to national military organisations. If one wishes to discern roots for exclusivist notions of national interest, for the atavism of the new imperialisms, for the social Darwinism which infected European great power relations, one had better begin not with the attitudes themselves but with the economic transformation of social structure and political power, and with the scientific transformation of the basis of military security. Thereafter the rationales of these changes in military contingency planning, alliances, imperialisms, and ultimately war between the great powers, extinguished from the international system before 1914 any remaining possibilities of a return to organic international relationships.

An accusatory finger has often been pointed at Clausewitz for the awful carnage of the First World War. This is unjustified. Clausewitz realised that no principle of moderation could be entered into the essence of war. Accepting this he understood that the military aim must always be set in some sense of defeating the enemy, and with some variation of maximum force concentrated at decisive points. But Clausewitz was adamant that war could be rational only when military aims and calculations of means were set in a context of political objectives whose purview must be the *status quo post*. Any war in which military aims dictated political objectives was irrational. Although he did not say this, for the institutional divide between civil and military authorities was not so great as it later became, it is implied in his work that military planning cannot provide a substitute form of rationality for careful estimation of political goals, risks and costs. Indeed as Bernard Brodie has skilfully argued in his analyses of the First World War and the war in Vietnam,[9] to permit an artificial separation of political object and military aim, to allow the second to dictate the latter or to be unclear in the former, is the abnegation of civil responsibility. Prima facie, adopting the arguments of Clausewitz, it is bound to lead to irrationality. The avoidance of irrationality requires that civil institutions always control military institutions by virtue, first, of social legitimacy, and secondly, a more organic understanding of national interest. In turn, however, this suggests that constitutive rationality as national interest functions best in an international system in which the terms of order are established and managed on an organic basis, on a basis of mutual recognition and legitimation of interests.

The evolution of the international system away from these conditions, and the effects of the technological revolution on military relations and attitudes towards security, adumbrates both the problems and relevance of applying the lessons of the classical strategists to the philosophical problems of arms control in an era of nuclear deterrence. This is more than an exercise in identifying the differences *between* contexts; it also points to the true measure of analytical difficulty arising from the dynamic transition from one context to the next, for it is here where the essence of change is located. The evolutionary nature of political, technical and cultural change, and the importance of theorising with these connections in mind, is reflected in Clausewitz. It also points to the coexistence of trends which might be seen as contradictory or even mutually transcending. Thus, in our own day, science has not displaced questions of ethics in matters of military power, and military power has not been used irresponsibly because it is ubiquitous. The historical tapestry of military and political relations since Clausewitz suggests an underlying tension between organic and structural views of international order, and between constitutive and demonstrative bases and manifestations of rationality. Arms control and science-based deterrence are contemporary battlegrounds of these views.

The Philosophy of Arms Control

The age of arms control began in the wake of the First World War. The League of Nations was dedicated to the abolition of power politics, and with that to ending the use of force and war as instruments of policy. The functionalist language and underlying precepts of the Charter saw international organisation, arbitration, and if necessary collective sanctions of an economic, or eventually military, kind as provisions which, with accumulated experience, would displace warfare by a quasi-legal international order. War was not declared to be abolished or illegal, but rather to be unnecessary. Disarmament became, for the first time, a goal of the system.

In particular provisions of the Treaty of Versailles, arms limitations were imposed upon Germany. The size and nature of the armed forces were prescribed. The High Seas Fleet would have been confiscated had it not been destroyed. The Ruhr was occupied; the Rhineland was demilitarised; and Germany lost territory, strategic raw materials, overseas colonies. Union with Austria was forbidden.

Over and above limitations imposed upon the belligerents, several of the victor states reached a naval arms limitation agreement in 1922. The Washington Treaty set ratios of capital ships and agreed common counting rules for displacement and armaments as a means of preventing naval arms races between sensitive nations in areas of strategic competition, the Mediterranean, Atlantic and Pacific in particular. There were further naval arms limitation talks in 1927, and an agreement between the United States, Britain and Japan in 1930.

All of these attempts to limit arms failed. Later efforts towards comprehensive disarmament did not proceed beyond unsuccessful international discussions, though a World Disarmament Conference met in 1932. The functionalist approach of the Charter failed because the terms of peace were not universally accepted. There were, indeed, more revisionist states than upholders of the *status quo*, and among these latter there was an absence of accord on the nature of security and the terms of order. It soon became clear that despite some successes. functionalism could only triumph over a return to balance of power politics with war as a rational instrument of states' policies given perfect initial conditions – an impossibly wishful condition. Particular attempts to minimise or stalemate the military power of the belligerents, and to keep Russia isolated in its own revolutionary context, also failed.

The system could not work without rehabilitation of the economies of the defeated. Given economic stabilisation, after much time and effort, there could be no valid moral or political objections to rehabilitation in other terms. Indeed, as Germany and Russia had already learned to conspire against the system from their isolated positions, the return to full legitimacy was prudent as well as judicious. As between legitimacy and sovereignty, there could only be pragmatic distinctions. France in particular sought to use her power and influence to sustain these but the use of security policies and, on occasion, direct military intervention to contain Germany was so inconsistent, so internally contradictory, that it worked no better in the context of the inter-war system in Europe than American containment of Russia, or Russia's doctrine of limited sovereignty in Eastern Europe, have worked since the onset of the Cold War. Prominent use of military factors in defence of the *status quo* proved to be costly, to generate tension, and ultimately to depend upon power relationships subject to erosion from domestic and transnational sources. Moreover, a *status quo* defined in military terms tends toward rigidity, and even if this can be sustained for long periods by the exercise of power or force or the construction of alliances, it is an obvious target for subtle and flexible manipulation – whether through direct arms competition or indirect means of subversion – by the revisionist state or states. Thus, when Germany rebuilt effective military forces, it did so with clear strategic objectives in view, and with novel doctrines and the newest of technologies. Similarly, Russia, Poland, Italy, Austria, and Hungary were all successfully induced to leave the ranks of the Containers of Germany, either by inducements of interest, or by intimidation of power. Under the direction of Hitler the general goal of German foreign policy, to revise the Versailles system and return Germany to its full plenitude of influence, became the basis of a compulsive strategy of indirection. In addition, it is arguable that the Anglo-Japanese Alliance which Britain relinquished as part of the cost of obtaining the 1922 Treaty, was a reassuring element of Japanese foreign policy which, had it been sustained, might have served to modify Japanese policy towards China and in the Pacific in the 1930s. Thus the first comprehensive attempts to alter the basis of international relations so as to limit the functions of military power and armaments, failed completely.

The founders of the United Nations were less idealistic when they came to frame the Charter. None the less, even the constitution of the great powers as a sort of Committee of Public Safety for the post-war international order was foredoomed by the gulf of disagreement which appeared between the United Nations and Soviet Union over the nature of that order. With the relegation of the Security Council to the periphery in the arbitration of international conflict there hardly existed any basis for international control of atomic energy; and with the advent of a thermonuclear arms race and war in Korea that stimulated a conventional – and later theatre nuclear – arms race in central Europe, it became clear that the foundation of any further attempts to limit arms could only be the maintenance of a stable balance of forces, and an overarching structure of deterrence. As between these two pillars of contemporary international security there were potential tensions, which alliance structures and inter-bloc arms limitations would be required to mitigate.

From the end of the Second World War there was a growing awareness among those concerned with strategic thought and planning, in the developed countries at least, that the character of international politics, the role of force, and possibly even the nature of war, had changed profoundly. This awareness did not come suddenly or all at once, but over a period of about a decade and a half there accumulated a body of new assumptions and theories concerning what was new, and what was not. It was assumed that nuclear weapons made a difference. Most strategists thought them to be neither revolutionary in international politics nor insignificant in military power, but somewhere in between: important and imponderable. From the ambiguous evidence of strategic analysis and Superpower crisis behaviour, an important set of theoretical insights into the nature of deterrence was derived. From this point onwards, too, a new logic of security began to be propounded: internationalist in essence, and morally liberal in its underlying objectives, it sought new terms for shared interests, stability, arms control, detente, limited war and crisis management. To a significant extent a functional structure of concepts, means of communication, and institutions, was built into Superpower relations and alliance structures. Some important attempts were made to extend arms control and detente into the developing world, particularly through non-proliferation arrangements and proposals for linkage; a brief flowering of international agreements to limit arms testing, development and deployment, mainly in non-crucial respects, and reflecting indirect Superpower objectives, occurred in the mid-1960s.

Yet all of this rested upon a conundrum. Many of those who in practical terms sought and implemented arms control measures believed philosophically that war had an enduring nature; that it had not been abolished, rendered obsolescent, or transformed by nuclear weapons and contemporary international politics. Some of the most respected civilian strategists, albeit that they were also sceptical of the nature of change, were students and interpreters of Clausewitz. Moreover, questions of whether if war occurred in the nuclear age, it could be limited; if it could be limited in practice, then according to what limits and how strongly these would operate; how limited war could be analysed, defined and provided with appropriate strategic and political rationales: these and other related issues were raised along with the prospect of a balance of terror. They were raised more acutely for the United States as the Superpower acting according to a doctrine of containment, and for NATO as an alliance facing an adverse balance of conventional military power in the European region. Thus, thinking about limited war in the nuclear age had to pursue unprecedented developments in strategic problems. Inevitably, perhaps, strategists often became both advocates of arms control and technicians of state policy. The cross currents frequently proved too strong for intellectual resolution.

Of the different possible ways of dealing with this issue, it seems that extended deterrence is as good as any. Extended deterrence can be thought of as the acceptance of *shared* risks in the defence of *shared* values. But what did this entail in practice once the two Superpowers had acquired thermonuclear capabilities, and in the light of the experience of Korea? This can only be understood in relation to the notion of central deterrence, the threat to implement the act of striking an aggressor, even though the consequence is to be struck back – destroyed – in the defence of *central* values. While rationally it may be 'better to be Red than dead', rationally also it is 'better to hang together, than to hang separately'. If, however, one can utter this latter belief of one's own society with rational equivocation, can one utter it of, or on behalf of, any other society? The crucial difference here is between constitutive rationality, which applies to central deterrence, and demonstrative rationality which – at least in the form of the problem of credibility – applies to extended deterrence. This clearly requires careful thinking about *conditions*, because the nature of political strategy is, as we have seen, essentially national, at least in the sense of being based upon a sovereign organisation of power, and a national structure of authority.

It is now possible to account for some of the philosophical attributes which arms control has evinced. Liberal and internationalist in essence, it has been informed by potent images of the peaceful resolution of great power conflict. One World, living in harmonious material growth, and a return to more flexible and co-operative international relations in Europe based on a balance of power without high levels of military confrontation. Beliefs of this kind, often associated with prescriptions for minimal strategic deterrence, denuclearisation, or at least de-emphasis of nuclear systems, in Europe, control of weapons proliferation to developing countries, bans or limits on weapons developments and testing, reflect an *organic* view of international relations, an emphasis upon the cultural forms of society and community, and a *constitutive* view of rationality and political choice.

None the less, much of the methodology and resultant analysis in the literature of arms control is philosophically conservative. Why is this so? In the first place it is easier to understand a cautious idealism and a resultant conservatism in the light of the failure of the League of Nations, of General and Complete Disarmament in both interwar and postwar circumstances, and of the failure of postwar attempts to provide for international control of nuclear energy. The Munich analogy rings true even in the minds of critics of 'hawks', whether they be American or European. Secondly, the institutionalisation of the Cold War and the Evolved Cold War on the basis of strategic deterrence appeared to recreate a paradox of the kind coined by Vegetius: 'si vis pacem, para bellum'. Although many who advocate arms control seek a resolution of this paradox, they admit its power. Clausewitz, the most respected of classical strategists, presents an ambiguous picture of force and war to the contemporary strategist who would prognosticate upon controlling military power. Without clear cut political objectives towards which nuclear weapons might be used, war in which their use would occur is irrational. None the less if war occurs, it remains a demonstratively rational military aim to bring to bear the maximum force, concentrated upon the centre of gravity of the opponent, within the context of political interests set by legitimate civil authority. As the parameters of nuclear deterrence remain intangible, the consequence of this distilled wisdom is to shift the onus of military power from *using* to *threatening* to use force. Simultaneously, this invites and perhaps obliges surrogate analysis of political objectives and military aims. This cannot be accomplished without ascribed rationality, but the result is a theoretical diplomacy of violence. Hence in the very act of serving to provide a legitimate complement to military contingency planning in conditions of continuous competition and deterrence, arms control confronts, and in a way serves to create, problems of strategy, and in particular problems of limited war.

It is in this connection that the rise of science has both served and damned the cause of arms control in its fundamentals: the limitation of military power to sensible proportions in keeping with the risks of war. If nuclear war is irrational yet possible, and if the setting of contingent military aims requires the invention of demonstratively rational but untestable political objectives for the opponent, then continuous scientific weapons development becomes the crucial solution to defining national security. Indeed in many ways it seems also to become the key to defining relations between civil and military authorities. Arms control tends to become arbitrational, little more than the litmus paper of the weapons laboratories. Hence, perhaps, why it appears not only to fail by obtaining conservative results from liberal ideals but also, at the same time, to function satisfactorily in the management of great power relations.

The structure of civil-military-industrial complexes and concomitant

'National Security States'[10] is attributable less to the nature of capitalism, or Soviet imperialism, or Superpower competition, than to the functional relationship which has grown up between military power in its connections with deterrence, and conceptions of national security in their connections with science and assessments of risk, chance, intentions and capabilities, as outlined earlier. In this complex structure of which it must be a part if it is to function, arms control has become inextricably linked with the *status quo*.

There is much here to suggest that it is deterrence which must be impeached for rendering arms control conservative. It would be mistaken, however, to conclude simply this. Rather, it is the entrammelling of deterrence with criteria of *demonstrative rationality* which appears at fault. The emergence, and early institutionalisation of deterrence in great power relations had relatively little to do with continuous weapons innovation or continuous military confrontation. The permanent interconnection of security, science, the military and arms control, came only with indications of instabilities in doctrines and force postures. What appears to explain then, why strategic competition is the requirement of credibly extending deterrence in circumstances of evolving nuclear balance between the Superpowers? Latterly, this could not be maintained within the Alliance without a doctrinal figment: NATO's variant of Flexible Response. But this doctrine bound the United States and its European allies into a close relationship which expressed their common interests and mutual security, and guaranteed that *because* of this arms control and high levels of military capability would be twin priorities. It was thought at the time that they could be complementary. Indeed, a certain complementarity was created so long as arms control negotiations were structured along functional and largely technical lines, but the cost of this was to atrophy any meaningful political dialogue about military power. In this triple connection (doctrine, capability and technicality), the requirements of extended deterrence bound arms control ever more closely into a conservative dilemma from which, whether in SALT, MBFR, other arms control arenas, or more recent assessments of European security and arms control, it has found no escape. Extended deterrence as it has evolved has been the kiss of death for any liberal philosophy of arms control.

Finally, then, the contemporary philosophy of arms control is inherently weak and contradictory. Its spirit has been liberal and its essence has been idealist rather than positivist. But its *methods*, for a mixture of good and bad reasons, have been conservative. Its positions on issues are entangled with dilemmas of theory and doctrine in alliance relations. Its objectives are remote from controlling military power to proportions and purposes set by reason. Thus objectives in their very prudence and empirical nature have become part of an intensely conservative policy of deterrence which ignores fundamental issues of military power. Economising upon military expenditure, stabilising arms postures, offering reassuring platitudes to restless publics, providing for crisis communication and, occasional rules of restraint, symbolising the importance of non-proliferation, and managing alliance interests: these are not unworthy in themselves. They do, however, require an intimate linkage to the underlying political bases of security if they are not to become self-serving rituals. Today arms control has no clear position to take on the requirements of deterrence, on regional nuclear systems or battlefield weapons in Europe, on the intimate connections between so-called Forward Based Systems and prospects for conventional force reductions, on the new military containment which appears to characterise American strategy, or upon the relationships between continuing strategic competition in the developed world and sources of instability in the developing world. This is because these issues lie outside the traditional purview and uneasy consensus of a conservative philosophy of deterrence, of which arms control is a part.

Conclusion

At present we are profoundly uncertain about the significance of having failed to control the spread of weapons of war during an era of unprecedented peacefulness in the international relations of the great powers and their allies. In part this uncertainty stems from the rate of development of new systems with war-fighting characteristics. In part it stems from the ambivalent state into which arms control has fallen. In part it stems from the increasingly unstable state of Superpower relations, which are both moving unsteadily towards a kind of strategic equality unmatched in history, and constantly threatening to be blown off that unsteady course by fundamental ideological antagonism, or by wars and crises in the world beyond.

The great fear – and who is able to deny it – is of a crisis which will change the balance of power psychologically even if it does not trigger war. Humiliation for either Superpower, as the Soviet Union was humiliated in the Cuba Missile Crisis, might be disastrous. Why is this? It is because the use of force is no longer about confrontations arising over largely demarcated commitments in and around Europe, testing the political resolve of great powers evidently restrained from military adventure, and deterred by the centrality of interests and nuclear forces. Today the use of force is about confrontations in grey areas of undoubted strategic significance in parts of the world beyond the ambit of demarcated commitments where restraint is not obviously prudent and where nuclear deterrence is not directly applicable, though limited nuclear use may be seductive. Yet it is these more likely contingencies which arms control in its connections with deterrence has treated as intellectually and practically secondary.

This is the fundamental problem of arms control: to prevent an over-reliance upon military power in its technological and allegedly deterrence-related forms from bringing about war in the developed world, or between the developed world countries; and indeed, to reduce present reliance upon military power wherever this can be said to obstruct natural political settlement. The difficulty in integrating these concerns lies in the epistemological problems outlined earlier with regard to the state of arms control and the evolving patterns of conflict.

The developed world exists ever more clearly as an island of deterrence in a sea of international conflict that closely follows classical models. Yet it is scarcely valid to assume that deterrence explains the absence of war between the Superpowers and their allies. The absence of war is better explained by reference to a range of social factors which are so structured together as to provide a context wherein military factors function to constrain political interest in the use of force, and political disincentives positively discriminate against the continuation of conflict by other means. This suggests that for 35 years now we have been living in a singularly fortunate era of arms control, but have quite perversely and falsely sought objectives for arms control outside the objective context in which arms have been controlled.

Arms control has assumed that the role of force has *changed*, and not the logic that force has *not* changed its nature, but only its place in the scheme of things. This half-logic explains the underlying suspicion with which advocacy of arms control has been viewed by the majority of military and political leaders. But it must also be concluded that it is the selfsame suspicion which fails to understand that absence of war cannot be explained only by presence of weapons and their emanating political effects, but must be explained also by self-restraint, not only on one side but on all sides.

It follows then that the objective of arms control is not simply to save money or to enhance stability, but precisely to ensure the preservation of this overall balance between policy and arms. Here is where arms control has failed, and been failed by others. It has been incorporated within the Evolved Cold War. That structure has controlled military power functionally. But it has shown a considerable disability to control the place of military factors in its international relations, or to respond to alarming signs of change and instability in the developing world. The idea that arms control could serve as a leading edge of the political control of force on a global level has failed. The quantity and technological sophistication of military forces deployed in the developed world, in the context of its underlying conflicts, is massively out of step with existing reserves of political restraint, and with the increasing historical understanding – and evidence – we now have of the origins and development of the Cold War.

To change the place of arms control within the Alliance and with regard to Europe is, in the first place, to change the way in which the control of arms is *thought* about, rather than just the objectives sought, the negotiating frameworks adapted, or the contexts of politics chosen. Although a realistic philosophy of arms control has existed for at least twenty years, and should not be abandoned, its premisses are in need of fundamental review. Arguing that arms control must operate within the existing strategic context of the developed world rather than the entire international system, that is must serve alliance management, that it must reinforce but never detract from deterrence, and that its proximate objective is to economise, may be prudent but is now inadequate.

What has changed, and what therefore makes reviewing arms control so crucial, is the existing strategic context, the conceptual and operational characteristics of deterrence, and the significance of military power in developing-developed world international relations. The Alliance will find it increasingly difficult to use arms control to respond to these changes. They require a fundamental revision of interests and a will to change. Arms control has failed its own best hopes, functioning palliatively within a remarkably stable – or should one say rigid – pattern of military power. It has, thereby, lost its relevance to problems of change, both with regard to ameliorative developments in political relations in the developed world, and the critical strategic links between developed world stable competition and developing world conflict. However the fault lies within ourselves, not within arms control as such.

Notes

1. Geoffrey Barraclough, *An Introduction to Contemporary History* (Penguin, Harmondsworth, 1964), p. 17.
2. Benjamin S. Lambeth, 'Soviet Strategic Conduct and the Prospects for Stability' in *The Future of Strategic Deterrence, Part II*, Adelphi Papers no. 161 (IISS, London, Autumn 1980), p. 27.
3. Ibid., p. 28.
4. Bernard Brodie, *Strategy in the Missile Age* (Princeton University Press, New Jersey, 1965), p. 276.
5. Bernard Brodie, *Escalation and the Nuclear Option* (Yale University Press, New Haven, 1966).
6. Malcolm W. Hoag, *Counterforce, Conventional Arms, and Confusion*, Rand Paper P-6485 (The RAND Corporation, Santa Monica, Cal., 1980), pp. 6-7.
7. Ibid., p. 11.
8. Michael Eliot Howard, *The Franco-Prussian War* (Collins, London, 1967).
9. Bernard Brodie, *War and Politics* (Macmillan, London, 1967), Chs. 1, 4, 5.
10. This concept is used to depict the domestic aspects of United States foreign military policy in the Cold War by Daniel Yergin in *Shattered Peace: The Origins of the Cold War and the National Security State* (Andre Deutsch, London, 1978). However, it was used by opponents of the Nixon Administration's defence policies in a work published in 1969, as Henry Kissinger notes in *The White House Years* (Weidenfeld and Nicolson and Michael Joseph, London, 1979), p. 199 and chapter note 4, p. 1479.

NOTES ON CONTRIBUTORS

Desmond Ball is a Research Associate at the Strategic and Defence Studies Centre at the Australian National University, Canberra. Part of this chapter was written when the author was at the International Institute for Strategic Studies, London.

Richard Burt is Director of Political-Military Affairs in the US Department of State. This chapter was written when the author was National Security Correspondent with The New York Times.

Paul Buteux is Associate Professor of Political Studies at the University of Manitoba, Canada.

Lawrence Freedman is Head of Policy Studies at the Royal Institute of International Affairs, London.

Colin S. Gray is Director of National Security Studies at the Hudson Institute, Croton-on-Hudson, New York.

Lawrence S. Hagen is a Defence Analyst with the Canadian Department of National Defence, Ottawa. He is also a doctoral candidate in International Relations at the London School of Economics and Political Science, where this volume was conceived and edited.

Pierre Hassner is Senior Research Associate at the Centre d'Etudes et de Recherches Internationales of the Fondation Nationale des Sciences Politiques, Paris.

Hugh Macdonald is Lecturer in the Department of International Relations at the London School of Economics and Political Science.

Robin Ranger is Associate Professor at St. Francis Xavier University, Nova Scotia, Canada. He is currently with the Defense and Strategic Studies Program at the University of Southern California, Los Angeles.

Lothar Ruehl is Deputy Government Spokesman for the Federal Republic of Germany. This chapter was written when the author was

Director of the Brussels office of Zweites Deutsches Fernsehen.

Helmut Sonnenfeldt is Guest Scholar at the Brookings Institution, Washington. He was formerly a Counsellor at the US Department of State.

Philip Windsor is Reader in the Department of International Relations at the London School of Economics and Political Science.

INDEX

ABMs (anti-ballistic missiles) 21, 22, 32, 47, 60, 61, 69
ABM Treaty 45, 53, 64, 73, 105, 108, 114, 187
ACDA 80, 106, 107
Afghanistan, invasion of 15, 37, 49, 56, 73, 97, 99, 170, 171, 174, 176, 182, 188, 192, 193, 203, 204
ALCMs (air-launched cruise missiles) 86, 87, 104
Algeria 203
Angola 15, 189, 200
anti-satellite systems 73, 122-3
Arab-Israeli wars 9, 15, 31, 145, 169, 200, 203, 205
arms control
  advantages of 70-3
  attitudes to 41-2
  bureaucracy and 106-7
  criteria for 22-5
  criticisms of 100-13
  defence planning and 67-70, 72, 76, 78, 104
  detente and 25, 28-30, 48-9, 56, 97
  deterrence and 27-8, 237, 239
  difficulties of 94-100, 109-10, 204, 205
  fallacies of 63-6
  historical background of 231-8
  international relations and 204
  military programmes and 71, 72, 197, 208-10
  negotiations, difficulties of 62-3
  philosophy of 240
  politics and 54-5, 75
  security and 27-40, 56-73, 94-115
  strategic stability and 43-8, 49, 59
  verification 62, 65
  Western misunderstanding of 110-13
  working of 110-11
  see also Europe, arms control in, MBFR, SALT
artillery shells 125
Austria 228-9, 233

B-1 bomber 103
B-52 bomber 61, 128, 130
B-61 bomber 125
Backfire bomber 23, 24, 66, 79, 87, 88, 91, 129, 146, 150, 154, 179
Bangladesh 203
Barraclough, Barry 198
Baruch Plan 67
Battlefield Nuclear Forces 86, 88
Beaufre, André 14
Belgium 50, 72, 81
Berlin 12, 13, 29, 171, 183
Bertram, Christoph 71, 94, 104, 112
Bohemia 228
Brandt, Willy 29, 35
Brezhnev, Leonid 49, 87, 177, 189, 190, 192, 198
Brodie, Bernard 108, 218-19, 230
Brown, Harold 103, 109, 135, 136
Bull, Hedley 67
Bundy, McGeorge 77, 178
Burt, Richard 106, 112

C_3 (command, control, communication) 134, 136, 137-9
Canada 189
Carter, President Jimmy (James Earl Jnr) 39, 50, 56, 59, 66, 67, 98, 106, 132, 135, 136, 170, 190
CBMs (confidence-building measures) 53, 59
CD (Conference on Disarmament) 90, 91
CDE (Conference on Disarmament in Europe) 82, 90, 91
Chatham House 103
China 133, 189, 200, 213, 233
Clausewitz, Karl von 9, 13, 200, 217, 221-6, 227, 230, 231, 234, 236
CND (Campaign for Nuclear Disarmament) 74
Cold War 9, 10, 11, 12, 13, 15, 20, 199-200, 214, 236, 240
COMECON 193
conflicts, regional 15, 16, 33, 59, 202
counterforce strikes 136-9

Crimean War 228
cruise missiles, arms control and 65-6
  see also ALCMs, GLCMs, LRCMs, SLCMs
CSCE (Conference on Security and Cooperation in Europe) 36
Cuba, 15, 98, 182, 200
  see also following entry
Cuban Missile Crisis 13, 14, 16, 21, 173, 238
Czechoslovakia 12, 14, 15-16, 19
  invasion of 28-9, 30, 173

De Gaulle, President Charles 29
Denmark 81, 229
detente 10-20, 188-9, 209

Egypt 192, 193, 200
Eighteen Nation Disarmament Conference 90-1
El Salvador 170, 173
engines, developments in 123, 125
enhanced radiation weapons 75, 88, 125-6, 146
Equivalent Countervalue 219-20
Ethiopia 200, 203
Europe, arms control in 36, 37-9, 74-92, 94, 99, 102-3, 113-15, 152
  see also NATO, Warsaw Pact

F-111 163
FB-111 bombers 92, 128, 130, 146
Fencer (SU-19) aircraft 147
First World War 9, 43, 230, 231
force, patterns of use of 199-201
Ford, President Gerald 98
France 78-82 *passim*, 88, 90, 114, 149, 170, 174, 177, 186, 228-9, 229, 232
Franco-Prussian War 9, 229
Freedman, Lawrence 22
fuel-air explosives 126
FUFO (fuel-fuzing option) bombs 125

Gelb, Leslie 62, 112
Germany 232, 233
  see also following entries
Germany, East 182-3
Germany, West 18, 34, 36, 48, 72, 81, 82, 88, 89, 155, 170, 171, 173, 175, 177, 180, 183, 186
Giscard d'Estaing, Valery 171, 174
GLCMs (ground-launched cruise missiles) 61, 79, 91, 104, 114, 148, 152, 161, 162, 163, 164
Gromyko, Andrei 36

Hassner, Pierre 20
Holland 50, 114
Honest John missiles 147
Hormuz, Straits of 171
Howard, Michael 110-11, 229
Hungary 192, 228, 233
Huntingdon, Samuel 173

ICBMs (intercontinental ballistic missiles) 21, 22, 23, 45, 47, 50, 124, 127-32, *passim*, 139, 146
  see also under names of types of
IISS 103, 112
India 200
Indian Ocean 56, 59, 66, 73, 202
Indonesia 200
Iran 73, 98, 169, 170, 172, 176, 182, 183, 192, 202, 203
  war with Iraq 168, 170, 200
Iraq *see previous entry*
Israel 193, 204
  see also Arab-Israeli wars
Italy 182, 233

Japan 182, 189, 233

Kennan, George 182
Kennedy Administration 13, 213
Kissinger, Henry 36, 57, 135, 146, 174, 178
Korea 12, 193, 200, 235
  South 169
Kristol, Irving 169

Lance missiles 146-7
laser weapons 122
League of Nations 231, 235
Lebanon 193, 200, 203
Legvold, Robert 171
Libya 200, 202
'linkage' 17-19, 30, 76
LRCMs (long-range cruise missiles) 86, 87, 88, 104

McCarthyism 213
McNamara, Robert 22, 29, 38, 44, 64, 146, 160, 220
MAD (mutual assured destruction) 13, 15, 19, 44, 221

Makins, Christopher 113
Malaysia 200
MBFR (Mutual and Balanced Force Reduction) 36, 47-8, 53, 62, 66, 72, 91, 110, 113, 207
Middle East 14, 30, 33, 38, 198
see also Arab-Israeli wars *and under names of countries*
military power
attitudes to 196-7
ideologies of 211-16
Minuteman missiles 57, 61, 124, 126, 127, 128, 130, 161
MIRVs (multiple independently targetable re-entry vehicles) 30, 45, 121, 206
missiles, guidance of 123-4, 125, 126, 164
Moscow 61
MRBMs 91
Mya-4 Bison bomber 129
M-X missiles 41, 61, 64, 67, 68, 69, 70, 72, 103, 104, 105, 124, 126-7, 130, 132-4

NATO
arms control and 74, 75
condition of 1979 37-8
continuity and discontinuity within 176-80
disunity in 75, 168-9, 173, 174, 175, 203, 204-5
extension of 193, 203
symmetry and asymmetry of 180-2
technology and 144-7 *passim*
theatre nuclear weapons and 37-8, 50, 71, 74, 81, 83, 89, 113-14, 147-51 *passim*, 152-65, 174, 175
vulnerabilities in 182-4
neutron bombs *see* enhanced radiation weapons
Nigeria 200
Nixon, President Richard 30, 102, 135
Norway 81
nuclear war, scenarios for 160-1, 207-8

oil 58, 176, 193, 195, 204

Pakistan 200
'parity' 22-5 *passim*, 46-8, 109
particle beam weapons 121, 122
Pershing missiles 49, 61, 72, 91, 92, 114, 148, 150, 152, 161, 162, 163-4, 181
Persian Gulf 56, 58, 192, 198, 202, 203
Poland 15, 19, 168, 170, 171, 173, 233
Polaris missile 65
Poniatowski, M. 174
Poseidon 124, 128, 130, 146, 161

QRA (quick reaction aircraft) 163

Radar 45, 134, 137
rapid deployment force 203
Reagan, President Ronald 39, 52, 70, 71, 74, 75, 76, 77, 80, 89, 170, 173, 182
Rose, François de 180-1
Ruehl, Lothar 113

SACEUR 161, 163, 205
SALT (Strategic Arms Limitation Talks)
cruise missiles and 64
detente and 17, 18, 19, 22
European security and 34, 179
nature of 20-1
politics and 29-31, 33, 37, 39, 98
provisions of 45
regional conflict and 32-3
strategic relations and 20-6
theatre nuclear weapons and 76-82 *passim,* 89
satellites 123, 138
see also anti-satellite systems
Schlesinger, James 23, 135, 145, 146
Schmidt, Helmut 34, 49, 77, 81, 83-4, 158, 170, 171, 182
SLBMs (submarine-launched ballistic missiles) 21, 47, 60, 128, 129, 130, 131, 146
SLCMs (sea-launched cruise missiles) 76, 79, 86, 87
Somalia 200, 203
Sonnenfeldt, Helmut 35
South Africa 200
SRBMs (short-range ballistic missiles) 76, 79
SS-4, SS-5 92, 148, 149
SS-11, 127, 128, 129
SS-12 91, 92
SS-13 127, 129

SS-16 124, 132
SS-17 60, 124, 127, 129, 131, 132
SS-18 60, 72, 121, 124, 127, 129, 131, 132
SS-19 60, 124, 127, 129, 131, 132
SS-20 24, 58, 61, 72, 76, 79, 88, 91, 114, 146, 147, 148, 149, 150, 154, 179, 182
SS-22 92
SSBNs 44, 60
SS-N-5, SS-N-6 129
SS-N-8 125, 129
SS-N-9/-12 76
SS-N-18 60, 129
strategic forces, balance of 57-9, 121-39, 190, 191, 205-6
submarines 44, 60, 121, 138, 149
Suez 169

Taiwan 201
Tanzania 200
technology
  strategic relations and 21, 25, 32, 65, 121-2, 126-9, 150, 206-7
  weapons and 13, 43, 122-6
Thatcher, Margaret 77-8
Titan missiles 128, 141
Tomohawk missile 49
Trident missile 61, 70, 81, 124-5, 128, 130
Trident submarine 103
Tu-16 150
Tu-22M bomber *see* Backfire bomber
Tu-95 Bear bomber 129
Turkey 58, 146, 182, 193
Typhoon missile 60

Uganda 200
Union of Soviet Socialist Republics
  air bases 140
  economic problems 193
  economic relations with West 187-8, 194-5
  military build-up 69, 79, 157, 191, 192
  military expenditure 190
  military programmes 190-3
  strategic doctrines 59-60
  theatre nuclear forces modernisation 76, 79, 89, 147
  United States of America, attitude to 215, 218
  see also Afghanistan, invasion of
United Kingdom 77-81 *passim*, 88, 114
United Nations 233
  Disarmament Conference 90-1
United States of America
  defence policy 68-70
  Europe, defence of by 34, 38, 77, 152-3, 162-3, 174
  military alert, 1973 15
  self-image 216, 217, 218
  strategic doctrines 13, 21, 25, 135-6, 153-4, 214, 219-21, 231
  USSR, attitude to 215, 216-18

V-bombers 163
Vietnam 14, 68, 190, 193, 200, 213
Vries, Klaus de 83

war, rationality and 226-7
wars 199-201
  see also under names of countries involved in
Warsaw Pact 19
  technology and 144-7 *passim*
  theatre nuclear weapons 147-51 *passim*, 191
  see also under names of member states
Washington Naval Conference 47
Washington Treaty 232
Wohlstetter, Albert 59, 178
World Disarmament Conference 232

Yugoslavia 12

Zimbabwe 200